COLOMBIA

Surviving the Long Conflict

Implications for the Elite Theory

Val Karanxha

HELIOS PUBLISHING
COLOMBIA: SURVIVING THE LONG CONFLICT
Implications for the Elite Theory

heliospublishing.us

ISBN: 978-1-7371175-4-4

Printed in the United States of America

Table of Contents

WORKS BY THE SAME AUTHOR

The History of Puerto Rico. Identity Formation and Nationalism. 2021

Identity in Spirituality. The Book About Self. 2022

Demystifying the Immigration Law: Implications for Practice. 2022

Interpreting the Role of Insurgency Inside of the Yugoslavian Wars. Article. 2023 (Saber and Scrolls)

Ideological Warfare: Interpreting Leftwing Insurgencies In Colombia. Article. 2024 (Saber and Scrolls)

The Legacy of Francophonie in Post-Colonial Africa. 2021

Yale New Haven Teacher Institute. Yale University

Writing from Images: Bringing the Latin American Heritage into the Classroom

INTRODUCTION

War and conflict in Colombia, for almost four decades, have cost the country thousands of lives, communities ravaged by violence, indigenous population and small farmers expelled from their land, and a divided country by rogue politics and impunity. Civil war is the last resort to solving failed policies by using violence and revolutionary means. It inflicts pain and suffering on millions affected by it. The conflict in Colombia was a mere case of "bellum omnium contra omnes." The state, the paramilitary, the drug cartels, and the left-wing guerrillas – were entangled in one of the most violent conflicts in Latin America from 1966 to 2016. The loss of innocent life was substantial. More than two hundred thousand people vanished, and many were never found. Impunity and government corruption reigned supreme. The country was decimated and balkanized in zones of influence between the guerrilla and the paramilitary. The state was fragile, and the threat of a possible collapse, at times, became a reality. The drug trafficking and the guerrilla warfare made Colombia a state and an institutional failure.

The public dissociated the conflict from its true nature – a struggle for the elite to remain in power. Instead, the public associated the conflict with drug cartels, cocaine empires, violence, and kidnapping, not to mention the coca labs in the Colombian jungle and the mined land in the countryside—the grim reality of the war almost brought the state on the verge of collapse. The conflict had many layers, including key players in domestic politics and international actors. The power struggle became intense – particularly in the 1990s. The fighting parties competed for power and territories inside the Colombian conundrum. They sought no other ways or solutions to appease the conflict except violence. The lack of willingness to end the conflict was mirrored in the weak attempts for a peace process between the state and the guerrillas. The right-wing militia often sabotaged these peace talks.

The violence seemed a sort of modus vivendi and had deep roots in people's mentality, and the country was often referred to as a necromancy. During the 1980s and 1990s, Colombia had the highest crime rate globally. It was a dark age. Images of Colombian fighters training in the jungle and the indigenous farmers working in the coca labs are still present in our collective memory.

The drug cartels were crucial players in the 1990s conflict but were not the only ones. There is more to the anecdote than the infamous drug trafficking and guerrilla warfare. The public has learned a great deal from documentaries and series yet lacks the complete panorama of the events.

Due to Colombia's importance in the global economy, the conflict risked paralyzing many parts of the agricultural sector and disrupting the supply chain. Thus, international actors were vested in the economy, especially in trading. [1] The international conglomerate and the fruit companies were paramount in the conflict.[2] In cooperation with wealthy Colombian landowners, the corporations encouraged the expulsion of small farmers and indigenous populations from the arable land. Not only did it destroy the small agrarian class, their crops, and livestock, but it also drove hundreds and thousands of people out of the countryside to the crowded urban areas in dire poverty.[3] The international conglomerates used paramilitary troops to secure their interests.

In 1994, Colombia exported 3 billion dollars in agricultural products. The conglomerates deepened their involvement in the agribusiness.

Colombia has always played an essential role in Latin American politics and has been a strong ally of the US. During the 1960s, a left-wing movement prompted insurgencies and a possible spillover of the Cuban Revolution. The cooperation between the US and the Colombian state intended to stop the spread of ideological warfare, keep the USSR's influence far from the Western Hemisphere, and avoid a second Cuban revolution. Ideologically, the left-wing insurgency sought to cause a revolution. Deterring a possible "Sandinista" movement in Colombia became a primary task for the state of Colombia.

US agencies such as Special Forces Operations SFO cooperated with the Colombian State to deter and dismantle the FARC guerrillas. [4] In the early 2000s, the priorities shifted with the spread of terrorism. Drug trafficking was not only sponsoring the insurgencies, but the funds risked being funneled into terrorist organizations.

Crime and Impunity

The conflict divided political actors into many camps, and few were ideologically opposed. Disclosed files of the Department of State revealed that the Colombian state outsourced its powers to the paramilitary to crush insurgencies and eliminate the cocaine kingpins. The Human Rights Watch has recorded many crimes not only committed by the paramilitary on innocent civilians, but it also has evidence of direct state involvement in committing violent acts. Going back to the 1980s, the political turmoil that plagued the country gave the Colombian state a tarnished name.

In his book "Rogue States" Noam Chomsky explains in detail the Colombian state's involvement in human rights abuses. It is worth mentioning that state-sponsored violence, directly or indirectly through the paramilitary, means the same amount of executions and extrajudicial killings.[5]

Paramilitary cooperated with the army and conducted illegal activities, laundering money or using drug money to finance their operations against the FARC - EP (Fuerzas Armadas Revolucionarias de Colombia) – Revolutionary Armed Forces of Colombia—People's Army.

The US Department of State disclosed that Chiquita cooperated with the AUC (Auto-Defensas Unididas de Colombia), the Colombian paramilitary, even after the US government moved them to the list of terrorist organizations. The paramilitaries were responsible for displacing small farmers from their property to expand the land needed for international agribusiness.

The conflict was much more complex. At times, the fighting parties conjoined with one another and then later became mortal enemies. The guerrilla ideological warfare, the paramilitary massacres, and the extrajudicial execution of Colombian citizens by the state were undoubtedly a sad case of the state gone awry. [6]

All the belligerent parties, including the guerrilla fighters, the paramilitary troops, the drug cartels, and the government forces, share responsibility for the human rights violations in Colombia. These actors committed crimes to advance politically and gain terrain in the ongoing conflict.[7] Violence in Colombia was pervasive, and people learned to survive it. More than two hundred thousand Colombians have died due to political violence since its beginning. In the last fifty years, Colombia has been stalled in a prolonged state of political violence, which has caused a steady decline in national security. [8]

The counterinsurgency operations were as harsh as the insurgency. Instead of supporting the areas affected by the conflict, the Colombian state adopted an aggressive approach toward people assumed to have different political views. It launched a witch-hunt campaign, especially in the urban areas where the military had taken over one-third of the territory. The country had never seen a political persecution of this dimension. The army went after syndicates, students, intellectuals, journalists, and clergy, in other words, against the subversive "population" suspected of sympathizing with the rebels. Under the National Security Doctrine, thousands of innocent civilians were tried in martial courts. In the first year of the presidency of Turbay Ayala (1978-1982), over sixty thousand civilians lost their liberty and were tortured brutally at the hands of interrogators.[9]

Therefore, the governmental structures were in this predicament. Many entities that operated in the shadow of the society were state-sponsored violent groups – and legitimized violence. The CONVIVIR was a private security group made up of militia and paramilitary. Widespread corruption in the government, involvement with the drug cartels and paramilitary activity, and financing of electoral campaigns with dirty money were factors that prompted the political instability in the 1980s and '90s. The governmental structures and agencies that had ties to paramilitary and narco-trafficking, whether local or national, included the National Police, the Colombian Congress, and presidential candidates.[10]

The Colombian conflict was multifaceted: The nature of the war was very delicate because it involved many vital players – domestic and foreign. There was a criminal network made of powerful drug cartels on one side. The (AUC- Autodefensas Unidas de Colombia), known as the paramilitary, were state-sponsored actors. Their primary role was to counterbalance the guerrillas on behalf of the right-wing political parties and the wealthy landowners. These structures operated in various zones, mainly in the Colombian countryside, such as Puerto Boyacá and the banks of Rio Magdalena. [11] The AUC was responsible for many crimes committed against the indigenous population and Colombian farmers who lost their land to a powerful agriculture conglomerate.[12] The state of Colombia faced a dire situation and an overwhelming task of combatting the guerrillas and the traffic of narcotics simultaneously. The increasing challenges to offset, at the same time, the insurgencies and the urban violence caused mainly by drug cartels forced the state to look for other solutions. The state needed to militarize, build its capacities, and create unique structures to combat the illicit organizations, whether guerrilla or drug cartels.

Institutional Crisis

Guerrilla warfare was mainly concentrated in the outskirts of Colombia. In contrast, political and institutional crimes occurred in metropolitan areas such as the capital, Bogota, where the political class and the elite lived and worked. The political class underestimated the threat coming from state failure and diminished its importance by shifting priorities from combatting the insurgency to eliminating drug cartels. At the time, drug cartels became an imminent threat to the Colombian political elite.

The illegal cash flow into the insurgency destabilized national security and the fragile national economy. Colombia was a model of a failed economy and institutional crisis. During the armed conflict, Colombia offered few opportunities for investors and international corporations. More importantly, a strategy to fight corruption in state institutions was needed to jump-start the economy and create a sense of security for domestic and foreign investments.

People became direct witnesses of massacres and displacements from their lands and homes. Thus, they lost trust in the institutions and the government. The drug cartel network was able to permeate state agencies. Once the state gets involved in illegal practices and crime, it immediately loses sovereignty and authority and undermines the rule of law.[13] Corrupt officials often became contingent on organized crime. As a result, impunity in Colombia became pervasive, and many victims of the violence never saw justice. In addition, as the conflict escalated, people turned away from the state and sought protection from the guerrillas. Young men and women flooded the guerrilla units to fight for what they lost in the war and seek revenge for impunity and missed justice. [14]

During the conflict, the judicial system disintegrated drastically. The law enforcement agencies were ineffective in dealing with the growing crime. Apart from the unrealistic caseloads, the judges systematically faced intimidation by organized crime. The drug cartels executed many judges who oversaw investigations of the illegal activities and homicides related to drug trafficking. [15] Judges who covered the illicit activities became dead men walking regardless of the protection offered by the state. Such was the case of Judge Gustavo Zuluaga Serna, assassinated in broad daylight by Escobar's assassin squad. [16] Death threats, kidnappings, and assassinations created a climate of fear in the judicial system and legislative process, culminating with the Justice Palace siege (Palacio de la Justicia) in Bogotá by M19.

The social disintegration and deterioration of the country caused pockets of vacuum power where the guerrillas and the paramilitary sought the opportunity to take immediate control. The guerrilla-controlled zones were out of state jurisdiction - de facto. Their fiscal system initially depended on the taxes FARC collected from the farmers and villagers in exchange for their security. The population under the guerrilla control was out of any governmental and state protection.[17] Guerrillas controlled strategic infrastructure and governed their territories inside the Colombian national territory. It was a shadow "state" run by insurgents and, for sure, a recipe for disaster. The FARC, at the time, controlled twenty-three percent of the national territory, meaning the state had lost territorial sovereignty to insurgents.

The FARC operated in rural Colombia, drastically weakened the state, and was a direct threat to the Colombian political elite. It threatened the state and its legitimacy. The FARC was an ideological enemy of the right-wing political groups (AUC), the agribusiness industry, and the economic oligarchy. Yet these entities, even though they only held economic power, could dictate inside politics and international relations. They ruled through the power elite and manipulated the state to fit their financial interests.

As a result of the lack of willingness to build a strong state, the growing tensions yield a Colombian state on the verge of collapse. The state had to find solutions outside its capacities to protect the national interests and the administration. First, strategically, the state could not finance a war against the insurgents, and secondly, it lacked the infrastructure and people to undertake such actions. As a result, the state decided to lean on powerful international actors and outsourced its powers to non-statal military units or militias within Colombia. Ideological and immediate financial interests prompted the paramilitary unit- to shift from private security to state-sponsored armed forces. Many of these units represented the Colombian oligarchy's and international corporations' interests.

The people of Colombia, especially in the rural areas, were caught in a tangle of rogue insurgencies with no way out. The clashes became increasingly destabilizing factors domestically and threatened regional security.

The insurgent forces were also competing for power with the Colombian state. Their immediate interests were to seize and control the territory and the drug routes. [18] The country's capital, Bogota, became the target of drug cartels and guerrillas who often carried out violent attacks by blowing up institution buildings, cars, and residential houses to achieve their political objectives. While Bogota remained a fragile political setting, the economic power was transferred to other parts of Colombia. Medellin became one of the strongest holds of the drug trafficking cartels. Valle del Cauca and Cali Cartel emerged in the international drug trafficking business, controlling other routes. Capital from illicit activities infiltrated various economic sectors, especially construction and housing.

The 1980s and 1990s were fraught with complexity in many aspects. The state had already lost much of its reputation for failing to handle the insurgencies and losing control over its territory. In addition, the Colombian state suffered a significant loss of legitimacy in international relations after the 1994 presidential election scandal.[19] The newly elected president Ernesto Samper, who allegedly had financed his electoral campaign with funds from drug cartels, caused a deadlock in the relationship with the US.[20] Samper's involvement with drug cartels was a political scandal that deepened the divide and increased violence and impunity.

Theoretical Framework

This study evolves around the elite's concept and role in preserving a state that would serve its immediate financial interest, a condition that translates into perpetual power. Hence, this book claims that Colombia fit the framework of a failed state during the conflict: It had lost jurisdiction or control over its national territory. The land was decimated by insurgent forces that competed for resources, territory, and power. While the Colombian state grappled to hold on to central authority, the local power was out of its reach and jurisdiction, weakening its control in the periphery and rural areas.

A paradox – perhaps is not explained as it should be: How do we account for this puzzle? The failed state indicators would possibly shed light on many unanswered questions and hypotheses that seek to explain the role of the Colombian state during the conflict. This study will attempt to explore and answer the following dilemmas and recalibrate the position of the power elite in times of conflict and disarray.

1. The Colombian state had the most prerequisites for a possible collapse. It had most of the indicators of the fragile state framework, yet it lingered for years and did not collapse. How do we account for this paradox in theory?
2. The Colombian elite that emerged in the mid-19th century embodied the same local character and lack of technocratic ability needed for the state to function. How does a political class perpetuate in a failing state without being replaced by another political class, whether by revolutionary or other reactionary means?
3. The fighting insurgencies (M19, the FARC, and ELN) never translated into transformational power or a full-scale revolution. After decades of fighting, the Colombian state successfully demobilized, diminished, and dismantled the insurgency. What factors contributed to their successful defeat?

Theoretical and Practical Approaches

FARC and ELN were entities that financed the war using drug trafficking money.

Nevertheless, no clear strategy was in place to bring about any change except to debilitate the state. I question the purpose of these entities. Were they only seeking to destroy the state and the political elite, or were they looking to form a shadow government that ran parallel with the state and coexisted as an alternative government? In this case, both these predicaments contradict the state's sovereignty framework. The study seeks to answer the question that we take for granted and believe at face value that what is good for the elite is suitable for everyone.

The conflict has prompted the interest of many scholars from military and political sciences backgrounds. Some arguments ignore the conflict's background and roots – Colombia's colonial past and social strata. Looking at the conflict through theoretical lenses, we can deduce that collapse is inevitable when the state simultaneously fights various insurgencies. Thus, theories are a bit rigid when framing the issues. Of course, there are many factors, among which is the critical role of the elite in preserving the state. Scholars from a military background emphasize the role that the US played in reducing the insurgency in number and operations. Plan Colombia was paramount to the state's survival. Ana Arjona, in her book Rebelocracy: Social Order in Colombian Civil War, reveals how insurgencies create parallel governments and establish some form of "normalcy" through two forms of social order, aliocracy or rebelocracy[21]

In the book titled Colombian Labyrinth: The Synergy of Drugs and Insurgency and Its Implications for Regional Stability, Angel Rabasa and Peter Chalk give an overview of the US involvement in Colombia using the rationale that Colombia was not only a narcotics problem; it was a threat to the national security of the United States.[22]

Mark Moyar, Hector Pagan, and Wil R. Griego, in the JSOU (Joint Special Operation University) Report, Persistent Engagement in Colombia, emphasize the role that the US played in Colombia during the conflict, revealing the urgency in combatting the insurgents. On the other hand – based on reports, the US Embassy in Colombia acknowledged the involvement of the Colombian Army with the paramilitary, their abuses, and the use of force toward civilians—various reports from US intelligence agencies – state that the paramilitary was responsible for the human rights abuse. [23]

The lack of transparency and the gross human rights abuses often hid behind the extrajudicial killings and non-proven guilty verdicts by the corrupt judicial system. Hence, the violence sponsored by the state and legitimized in the hands of private militia and paramilitary left a scar on people's lives and their memories.

Colombians have a different perspective and collective memory of the conflict. To many, it was very personal; to others, it was a prolonged process of pain and suffering without any closure. Even though the parties involved in the conflict signed the peace accord a few years ago, many Colombians impacted by the war have yet to make peace with themselves. The human tragedy changed people's perspectives about life. Nevertheless, the work written about Colombia by Colombians seems to have a common denominator – the culture of violence (la Cultura de la Violencia) and the power elite in the last seventy years at least.

They regard the conflict as a clash between the political elites and the insurgency that once belonged to the same political class. The rebellions in Colombia follow the same pattern. Both entities, the FARC and M19, were created as extreme factions out of the liberal parties, lacking the willingness to establish a Western political culture and values.

The theories seeking to explain the Colombian conflict are more complex than thought. The conflict itself was new and evolved over a long period. The geography, the economy, the state politics, and the elite determined how actors handled the conflict and what their interests were. Trends were very particular to the conflict, and the dynamics were specific to the country's geographical features and cultural behavior patterns. The mountainous terrain and the jungle favored the guerrillas and the paramilitary. The terrain gave them a considerable advantage in avoiding the state presence in the zones where they operated and conducted illicit activities. The agricultural roots of the conflict had their particularities. They determined the war's scope and magnitude in the zones where guerrillas and the paramilitary operated.

However, was the guerrilla war enough to have been called a revolution? The prolonged timeline changed its revolutionary physiognomy and priorities. After various actors enter the scene, sustainability is impossible to contain for more than a three-decade conflict. The parties involved in the conflict came from the rural areas and the political groups supporting the interests of wealthy landowners.

Later, more actors entered the conflict, including the drug cartels and the Colombian state. What started as an ideological war between the conservatives and liberals escalated into an armed conflict. Thus, when the state sought a solution to the problem outside of its capacities, the scope of the conflict had already expanded. With the conflict escalating, ideological warfare became a secondary influence in the political impasse. The parties fought for survival and control over the territories. The increasing tensions revealed the ineffectiveness and, therefore, the weakness of the Colombian state in front of spiteful forces. Nevertheless, Colombia remained a cash crop for international companies and the Colombian oligarchy. It is paramount to understand that the ruling class, whether the elite or any other regime with political authority over the country, will not surrender any political or economic power to the working or agrarian class.

This dilemma is the premise of this book. How do we approach the Colombian conflict? In conclusion, in this study, I will use two main theories to analyze the conflict that will explain why certain failed states do not collapse but return to their fragile state status. These two prerequisites are:

1. The failed state framework
2. The power elite and relations within the group

There is a pattern of political behaviors and political violence. They usually occur when entities with a political purpose use unconventional means to achieve a political objective. In the case of Colombia, these violent means were aimed at the Colombian government to pressure it to act on political and economic issues. Scholars, whether Western or Latin American, agree that the 1980s and 1990s marked a period of political and social unrest in Colombia, where the state had no control over its jurisdiction due to its inability to maintain sovereignty over the territory.

For example, FARC rose as the key player in the conflict due to failed agrarian reforms in the mid-20th century. FARC sought Marxism as the only way to redistribute wealth and end class warfare. Thus, the revolution was the only means to solve the ongoing struggle. Yet, we see a paradox that is typical of Colombia. A failed state that is unable to control a devastating war within its territory, yet it continued to overlook the conflict and preserve at face value a democracy with a pluralist system, three branches of the government, a separation of power between national and local governance as well as democratic, free elections.

Whether the conflict was contained or spread out to other regions, the broad perception is that the state of Colombia drove its citizens to an armed conflict and state-sponsored violence.

Theories of state tell a different account. The state must guarantee its own citizens' security – primarily. The monopoly over means of violence – was a case fraught with abuses and human rights violations. Colombia is a case of state failure because it functioned as a political unity with a democratic system yet was not able to control the territories that the guerrillas and other terrorist groups controlled. A state run by the political elite can control so much, yet it has failed to guarantee its people's security. The state no longer functions as an institution but as an instrument to preserve elite power.

To define the "health" of the state, we must look at specific indicators. Thus, evaluating Colombia's socio-political situation, one can deduce that the conflict reached new levels in the late 1980s and early 1990s. The state was fighting more than one insurgency; it battled organized crime and drug trafficking, which had spread to guerrilla and paramilitary groups. The country's economy was in shambles; informality and inflation were sky-high. The cash flow into the insurgency and the paramilitary brought more violence. Such high levels of corruption and instability indicated that Colombia was headed for a state collapse in all indicators, including state institutions. Yet the Colombian state did not collapse even after the M19 took over the Justice Palace in 1985. That was a defining moment that shaped the domestic policy in Colombia. Seizing the Justice Palace was a wake-up call, and the need to use the military arsenal to protect the institutions became pivotal in preserving the republic and the state. The Colombian government had to take drastic measures to secure a grip on Bogota's institutions and administration. While the capital became a priority in the fight against the rebels, rural Colombia remained fragmented and decimated.

Hence, how did Colombia avoid a total collapse? In the later chapters, I will explain how and why the political elite in Colombia—regardless of their differences in political ideologies—had to protect their existence and perpetuate their rule at any cost. While the elite maintained power in the metropolis, the insurgencies and paramilitaries operated in the rural and peripheral areas. It worked as an informal power-sharing agreement.

To overcome the crisis, the state divided priorities among actors – governmental agencies, law enforcement, the military, and the private militia. In the early 1990s, the Colombian state focused its efforts on the war against narco-trafficking since it became the sworn enemy of the elite and, therefore, the state. In the periphery, while the Colombian army outsourced its operations against FARC to the paramilitary groups – the Unites States also became part of the Colombian domestic conflict; it provided financial aid and military assistance [24] before and after "Plan Colombia."

The presence of the United States in Colombia – which I will explain in the later chapters- had to do with a few factors that governed international politics and the United States policy in Latin America. Colombia was willing to cooperate, and the central ideological threat to the United States – communism- was rising in many Latin American countries. The assistance from the US released some of the pressure the Colombian state faced domestically.

Indicators of a Fragile State

Security Apparatus and Security Threats to the State:

The theory suggests that the indicators of security threats, such as bombing, battlefield-related deaths, mutinies, coups, and terrorism, are prerequisites for a fragile state. The security apparatus also considers organized crime and criminal activity. The key to this indicator is that to combat crime and insurgency; the state takes extended measures such as mobilizing paramilitary or militia groups. In fact, in Colombia, this policy turned out to be disastrous. It legitimized the use of force and transferred the monopoly over the means of violence from the state to private groups.[25]

In other instances, the security apparatus may include a “deep state” that consists of secret intelligence units or other non-structured security forces that serve the interests of a political leader or clique. Finally, the security apparatus indicator will consider armed resistance to a governing authority. Some of the factors that impact the security apparatus have to do with the manifestation of violent uprisings and insurgencies, the proliferation of independent militias, vigilantes, or mercenary groups that challenge the state’s monopoly over the use of force.[26]

Factionalized elites:

The factionalized elites comprise the fragmentation of state institutions based on clans, political opponent parties, ruling elites, ethnic, regional, religious minorities, or majorities. In a broader sense, the factionalized elites in a fragile state represent the power struggle and competition for power and territories. Where the voting system exists as a form of free will to elect officials, they will factor in on how the power transits from one group to the other; otherwise, the group who holds power will continue to take control of the state. [27]

Thus, the critical point of the security apparatus and factionalized elites as indicators of a failed and fragile state are backed by a series of assumptions. Some of these assumptions consist of a paramilitary force and that of guerrilla or rebel insurgency; both can be instruments of elite factions. In addition, it is presumed that a fragile or failed state only partially controls the national territory. Most adamantly, in a failed state, the relationship between the citizenry and their security is characterized by violating human rights and state-sponsored violence. The police are characterized by unprofessionalism, corruption, and brutality. Lastly, the arms, weapons, and ammunition are present and available.[28]

Group Grievance:

According to the FSI methodology, the group grievance indicator focuses on divisions and schisms between different social groups—particularly divisions based on social or political characteristics—and their role in accessing services or resources and their inclusion in the political process.[29]

This indicator closely relates to the country's historical background or specific groups. It reveals the past and the injustices committed by the state or other political or social groups. Thus, going back in history, group grievance may shed light on how opposing political groups were formed, how events unfolded, and how conflicts shaped these groups and their place in society. [30]

Referencing history and historical events might clarify how and what crimes were committed during conflicts and what communities were victims of violence. Also, group grievances help the public understand the historical process of whether they were the target of state-sponsored violence. Group grievances point out instances when rights were denied or if the group has suffered persecution by the state or other political or ethnic groups. Above all, a group aggrieves when apparent impunity from the dominant or persecuting groups is prevalent. [31]

State legitimacy:

Based on the fragile state indicators, state legitimacy relates to the relationship between the state and its citizens. The state legitimacy indicator assesses the level of trust between the citizens and the state and evaluates the confidence in the institution of the rule of law. The state legitimacy indicator reveals how people manifest their opposing views and the right to protest against the government through civil disobedience or armed insurgency.

Though state legitimacy does not necessarily indicate democratic governance, it considers the integrity of elections when they occur. Furthermore, it considers whether the elections have been flawed or boycotted and if the peaceful power transfer is present during political transitions. The same indicator is used in case there is an absence of democratic elections and the degree to which the government is representative of the population it governs. Furthermore, the indicator considers the government's openness to transparency and accountability and the levels of corruption, profiteering, marginalizing, and persecuting opposition groups. Lastly, the indicator finds the ability of a state to exercise essential functions, such as the ability to collect taxes. [32]

Thus, to move my argument forward, I must justify that Colombia was a failed state by definition, and as such, it failed its citizens. The state of Colombia flopped its essential functions as a political and legal unit. It did not protect the citizens by providing physical and economic security. More importantly, the monopoly over the use of force resulted in abuses and violations of human rights by the Colombian military against segments of the population. The displaced population who joined the FARC and ELN saw the state as the enemy. Entire regions sought security provided by the rebels and declared themselves as *zonas guerrillas*.

Secondly, Colombia fits the definition of a failed state as it did not control the national territory. As a result, it did not have jurisdiction in the FARC-occupied zones. As time went by, the guerrilla took over vast rural areas and planted mines to keep the paramilitary and the state out. The mine-infested lands are still a risk to farmers and other rural populations dwelling in the regions.

Thirdly, Colombia sought to solve the domestic crisis by using the US help and assistance to keep the elite in power. The US's role in Colombia changed the dynamics of domestic and international policies and boosted the state's confidence in combating insurgencies and organized crime. [33]

Colombia fit the failed state framework because it used violence to solve a political crisis by engaging and allowing paramilitary troops to wage a dirty war against small farmers, the indigenous and rural population they alleged to sympathize with the FARC.

Colombian state failed to protect its citizens from pillage, expulsion, loss of property, and property rights, but most of all, it deprived their right to life and liberty.[34]

The Power Elite

There is a correlation between power and the elite. It is a top-down power concept that, even though it has evolved and has taken more modern features, is still pertinent and embodies strong characteristics of a society run by a ruling class. Regardless of various definitions and terminology changes, the core argument in the elite theory remains: Who has the power, and who does the power benefit? The core fundamentals of elite theory date earlier than Machiavelli, yet they have become relevant with Pareto and Mosca. Hence, the definitions change as the elite, besides generalized behavior patterns, also embody local and regional features. The theory presumes that society is always divided into two distinct groups: the ruling class, or the oligarchy, and the ruled, which are the people. Early arguments point out that the etymology of the word elite originates from eligere (Latin), the chosen ones, a definition by Vilfredo Pareto that embodies a positive connotation as opposed to the aristocracy. It is a ruling class by social position and not a necessary meritocracy. 35

The second definition is a more modern term, which implies that the elite is a group of people at the top of society, the political cupola, and controls the institutions. Elite's theory originates most clearly in the writings of Gaetano Mosca (1858- 1941), Vilfredo Pareto (1848-1923), and Robert Michels (1876-1936). Mosca emphasized how tiny minorities out-organize and outwit large majorities, adding that "political classes" – Mosca's term for political elites – usually have a "certain material, intellectual, or even moral superiority" over those they govern (1923/1939, 51). Pareto postulated that elites would be the most talented and deserving individuals in a society with unrestricted social mobility. Yet, in actual communities, elites are those most adept at using the two modes of political rule, force, and persuasion, and usually enjoy essential advantages such as inherited wealth and family connections (1915/1935, paras. 2031-34, 2051). Michels deemed elites ("oligarchies") needing large organizations for leaders and experts to operate efficiently. As these individuals gain control of funds, information flows, and promotions, power becomes concentrated in their hands (1915/1962; see Linz, 2006). They emphasize the inescapability and also relative autonomy of elites. [36]

Roberto Michels argued that the idea of the elite rests in its ability to organize itself as a group with shared interests against the disorganized majority. At the same time, Mosca's theory was modern and not based on ideological differences between the ruled and ruling classes. [37] He explained through methodology and observations that the political class he called the elite was a homogeneous group with a common interest at the top of the social hierarchy necessary for the state to function. Pareto, who was a senator, focused on elite dynamics. Pareto referred to the peak of the hierarchy as the home of the elite or the aristocracy, yet he mentioned that aristocracies do not last, nor do the elites; they move and are replaced by other elitist groups, and so forth. Pareto introduced the idea of multiple elites and how they merge into society. He distinguished between the economic elite, the intellectual elite, and the political elite. [38]

Regardless of the similarity between the scholarly interpretation of the elite, the critical point of the theory leaned toward a leadership and organization point of view more so than a political definition. In the mid-1900s, with the shift toward a society where "organizations" function as the hub of the systems and are pivotal for them to exist, the elite took a managerial role in society. [39]For example, James Burnham's book, *The Managerial Revolution: What Is Happening in the World*, shows this shift in how we understand the elite. In his view, Burnham leaves aside the Machiavellian school of thought and argues that the dominant political class, once made of financiers and owners, has now evolved into a class of professionals who needed to man the workforce. [40]

Although this analogy is more rhetorical than tangible, Burnham refers to a class of technocrats who are practical in keeping the system running. Burnham combined the analysis of Mosca and Pareto with the Marxist approach, concluding that the control and not the property of the means of production ensures access to power. Thus, in his theory, Burnham argues that it is about who controls the system and not necessarily who has control over the property and the means of production – which seems to be the case of the new American democracy instead of being ruled by the old "aristocracy" as Pareto and Mosca reason.[41]

Another theoretical view of the elite theory – is attributed to Roberto Michels and his perception of the oligarchy. Michels, who was a political activist, analyzed the mass parties and in his influential work Zur Soziologie des Parteiwesens in der modernen Demokratie: Untersuchungen über die oligarchischen Tendenzen des Gruppenlebens, Michels argues that the parties - unlike Pareto's aristocracy – are run by an influential minority that rules over the majority. He called this grouping – oligarchy. According to Michels, the oligarchy is indispensable for the organization's survival – in this case, the state.[42]

Local elites, on the other hand, are a bit more complex. Two arguments contradict one another; one is offered by Hunter (1953) and the other by Dahl (1961). Hunter points out that we see a monolithic elite at the local level that manages all the power. In contrast, Dahl underlines that elites are broken into groups that compete for power even at the regional level. [43]

Dahl's view is well illustrated in countries where the local elites operate with little regard for democratic processes. Such power struggles mirror how violence and conflict occur on the periphery. The regional elites also have a stronghold on the local police, determine outcomes, and pressure the central government about policies that would protect assets, properties, and wealth.

Divided elites and their loyalties: The demise of Grand Colombia

To understand the rationale behind the tumultuous politics and the state in Colombia, one must go to the root of the problem. The culture of violence did not appear in the mid-1900s, nor did the bloodshed of the 1980s. Sectarianism and the deep divide between the political parties and factions date back to the beginning of the Colombian state under the leadership of Bolivar.

The war of independence united the elites in the viceroyalty of Nueva Granada with the intent to throw the Spaniards out of the southern continent. Still, once the Spaniards were out, the division of elite clusters in Santa Fé de Bogota and Caracas started to show, revealing the cracks and the deep divide. A united Grand Colombia under Bolivar was a utopia for the Colombian elite. The schism, the intolerance, the calamities, and the uprisings or coups were deeply rooted in the newly emerged political elite and its mentality. Of course, after the liberation of New Granada from Spain, Creole elites started to form their regional identities. No longer was Bolivar the leader of the new Creole pioneers—the first failed attempts to build a

Colombian state date with the war for independence of Nueva Granada. The twenty-year quest for independence from Spain ended with many former viceroyalties wondering how to self-govern their territory. Amidst poverty, disasters, blood, and suffering that the royal army left while leaving Nueva Granada, the liberators had to face the task of building a state from shambles.

The emergence of the new ruling class in Gran Colombia had its complexities. The fragmentation of the political elite emerging from independence and the inability to assume leadership to build the nation posed a challenge. The new ruling class faced the challenge of administering the territory, building a treasurer, and completing basic tasks such as governing locally.[44] Local governance became a daunting task, the treasury was empty, and the laws that remained from the Royal Crown would become the fundamentals of governing in the first years. Moreover, Bolivar's dreams of universalism and human rights were overshadowed by his efforts to grab power and prevent a possible disunion of Gran Colombia.

During colonial mercantilism, creoles feared the commercial laws set by the monarchy, and the attitude toward the highest-ranking officials was somewhat sycophant, avoiding possible clashes and retaliation from the harsh royal system. As a result, Creoles avoided problems with the crown and the appointed judges. The trading system was very rigid, and minor violations would cause the merchandise to be blocked at the harbor and eventually miss the shipping deadline, even in the weighting system. Like the rest of the Spanish colonies, Nueva Granada had a system of strict guidelines on export and import. It meant a financial disaster for wealthy Creoles if the merchandise missed the ship to Europe. Thus, the colonial system shaped a wealthy class with pure economic interests and no governing experience or institutional culture. The new ruling class was the same wealthy landowners, enslavers, and merchants with significant political ambition and a narrow vision for the newly created states. The new Creole elite needed more experience in many areas, but the most crucial factor to the demise of Grand Colombia was regionalism and the failure to share power.

Bolivar's Universalism vs. Creole Elite Regionalism

"Colombian political history is characterized by the persistence of a deep-rooted local(ist) particularism that originated in colonial society and that, since the dawn of the "national" period, has been considered one of the most formidable obstacles to political centralization and national integration."[45]

Colombian elite can be described as one of the most solid political classes that, regardless of the ideological differences, held on to power during the conflict. Moreover, as an informal organization, it resisted threats from outside, especially from the drug cartels and the insurgencies. From the above definitions, a few apply theoretically in this case, yet going back to its conception as one of the most potent classes after the Revolutionary Wars headed by Bolivar, it had shown that this grouping was made primarily by the Creole elite that emerged during and after the war.

The primary source of wealth for the local elites depended heavily on land development, production of crops, and cattle. Culturing products specific to the climate favored the production of fruits and later prompted large international corporations and conglomerates to produce and export tropical fruits. During the 19th century, the local elites strengthened their economic position by exploiting production, exports, and trading. Regardless of the development of agriculture, the progress was modest. Nevertheless, in the second half of the 19th century, we began to see the creation of the petite bourgeoisie - a class of merchants whose trade depended on the production of tobacco and coffee.[46]

As I quoted at the beginning of the elite theory, I must reinforce Ospina's argument about the nature of the

Colombian elite. It has the features and characteristics of a local elite and operates based on its interests – with little regard for the majority. Also, modern elite theorists agree that this stratum (the elite) comprises technocrats and bureaucrats for the organization to function. In the case of the Colombian elite, the institutional culture was lacking, and no unified vision existed. Thus, there was a limited number of technocrats in the state apparatus. In the last 70 years, at least since the Korean War, Colombia has invested in military power and has kept a judicial system that would only work for the elite. For decades, the power-sharing agreement between liberals and conservatives reveals that this political class rules over the majority and compromises when necessary.

Going back to the emergence of the Colombian elite – it is worth mentioning that calamity attempts to overthrow Bolivar were prompted by the local elites headed by General Santander. Bolivar's Universalism was not only his idea of building a republic based on human rights; it was mainly a call to unite in front of the rising North American superiority. Santander rejected Bolivar's universalism, and with the support of the new ruling class after Bolivar's death, Gran Colombia was dissolved into various states.

Upon independence from Spain in 1821, Colombia passed various policies on land distribution nationwide. Yet many reforms had failed to produce the desired results and were designed based on the needs of large farms and cattle ranchers instead of taking a more upward approach for the landless peasants and poor farmers. Colombia has one of the highest levels of land inequality globally, with significant disparities in land laws and policies favoring wealthy landowners and cattle ranchers.

The land concentrated in the hands of the powerful landowners increases their efforts to secure a grip on the property and object to agrarian reforms on land distribution. Thus, their economic status grants them disproportionate power in the local communities. For example, this class of agrarian oligarchy sabotages efforts for equitable agrarian reform using their financial and social power. The rural elite will influence political participation by decreasing competition and increasing political party participation. By discouraging competition, the rural elite imposes control over local politics and influences representation in Congress.

The final objective of these local elites is to influence national policies on land and land distribution.[47]

Structure of the book

The book attempts to summarize the conflict in Colombia objectively and impartially, focusing on the state, political pragmatism, and elite power but, more importantly, solving a puzzle that can add a grain of contribution to the failed state theory. The format is straightforward and covers critical actors during the conflict. These major political events shaped the war and strategies used by the Colombian government to prevent a state collapse.

The first chapter, "The Roots of the Conflict," covers the conflict's genesis, emphasizing the elite theory's components, such as the local elites' mentality and the lack of a techno-bureaucratic system to facilitate the state apparatus. It shows an absence of state presence, institutional culture, and the inexperience of building a modern state. Chapter 1 argues that once the elites compete for power, the conflict will spill over to the rest of the country and take an unprecedented turn of events.

The second chapter, "Left-wing Ideology and Warfare," explains the theory behind the agricultural revolution, class analysis, and the search for solutions to the puzzle. It also analyzes how insurgencies arose and describes their ideology and structure.

Chapter three, "Colombian Paramilitarism," focuses on the conflict's paramilitary and their strategies to combat the insurgency. I present a timeline of the paramilitary rise, the national security doctrine, and the ideological grounds of the right-wing militia. I also explain the involvement of the AUC and various paramilitary groups with the Colombian government, the violation of human rights, and the crimes committed toward the rural and indigenous population in

Colombia. Lastly, I explain the political implications of international conglomerates in Colombia and their involvement with the paramilitary.

Chapter four, “Cartels Against the Ruling Elite,” describes how drug trafficking, particularly the Medellin Cartel, impacted the conflict as one of the critical political actors involved. Regardless of the popularity of the argument on an international scale, I present a different view. I have analyzed the cartel from an illegal organizational point of view, including its structure, operation, threats, and weaknesses that eventually led to its demise.

An essential role in this chapter is the presence of the United States in combatting narcotrafficking at an international level.

Chapter five, “Militarizing the State,” covers the role of the United States in Colombia and the region during the conflict. I have explained how the United States assistance changed the nature of the conflict and placed the state of Colombia from a vulnerable stage to an upper-hand position in combatting the insurgency and the narco-traffickers. This chapter also explains that the United States' presence, whether through financial support from Washington or otherwise, was crucial and helped the state survive.

Chapter six, “Challenging the Power Elite,” focuses on the theoretical application of the failed state in practice.

First, I explain the commonalities of a functioning state with a state nearing collapse, the relationship between the elite, and its role in preserving a state that works for the elite, not the people. Then, in this chapter, I present the political implications of the failed state of Colombia and its particularity. Finally, I conclude by emphasizing the failure of the elite to modernize and work in favor of the state. Instead, the elite used the state as an instrument to obtain power.

In the book's conclusion, I bring the puzzle pieces together, analyzing the nature of the Colombian elite in depth. I have used the elitist theory to highlight its political behavior, loyalties, and quest to control and survive beyond the conflict. I give an overview of Colombia and briefly describe the war's impact on essential population segments.

Delimitations

The book cannot and will not answer all the questions about the protracted conflict. It is not a historical book nor a narrative. The book does not discriminate against any of the fighting actors, nor does it take sides. It tries to build a theoretical framework about states that seek to outsource their powers to other non-structural actors to prevent a possible collapse. Furthermore, it aims to explain power relations among the elites or within the same elite to preserve the state and its legitimacy.

The Roots of the Conflict

Elites in Colombia have always competed for power, land, and capital, whether human or financial. As a ruling class, the elite emerged from the most ambitious political circles in 1819 after the viceroyalty of Nueva Granada declared independence from Spain. This grouping was made mainly of wealthy Creoles who were either directly vested in the war, supported the war, or fought along Bolivar. Many of them owned vast land and property and enslaved people—one of the challenges at the time in a quasi-independent South America was governing and nation-state-building. State failure in the newly created republics was not a new phenomenon. Elites often clashed over resources and prestige in the new political discourse. Soon, they developed their character as local, agricultural, and political power groupings. Juan F. Roman[48] argues that the two ruling parties (the Liberal Party and the Conservative Party) have much to say about Colombia's political conflict. He mentions – "When moderate factions ran the parties, the elite seemed to cooperate, as opposed to when extremists took control of the parties, the conflict escalated and the violence erupted." The conflict fueled by extremists sought the elimination of the moderates from the political discourse. [49]

In 1820, the Bolivar's army returned home to a free Grand Colombia, and the promise of owning land property was finally a reality. By the 1850s, the settlers, poor farmers, and indigenous people occupied about two or three acres of public land. At the time, the Colombian government encouraged the expansion by giving farmers concessions for planting tobacco, cotton, coffee, and other profiting crops. The wealthy proprietors lacked the human resources to work on their lands and sought the takeover of the farmers as the only opportunity to secure labor. Thus, the wealthy ranchers could procure land and labor by stripping farmers of their land. [50]

The expulsion and the expropriation of the farmers and the indigenous people from the land are known arguments in Marx's primitive accumulation,[51] and perhaps following his view, we can see the roots of the conflict and the core of the problem. In 1914 -1918, during the Concha administration, the indigenous population started a movement by Manuel Quintin Lame to recuperate the land appropriated by wealthy landowners and revive the indigenous communities. However, the expulsion of farmers and indigenous people from their land caused dissatisfaction and a widespread rebellion. Manuel Quintin Lame- the indigenous leader - who had spent years in prison organized a resistance movement followed by protesting that ended up in fire and blood. The military later pacified the uproar over land. Hence, it punished many protesters using the penal code for rebellion and sedition attempts.

In the 20th century, the Colombian agricultural business marked a period of prosperity and abundance in coffee export. It generated lucrative business for the dominant class that depended on land exploitation. The new Colombian oligarchy sought to expand the business by appropriating land occupied by squatters and small farmers. The coffee business put pressure on small farmers and caused dire economic situations for the economically disadvantaged. They faced two possible solutions: either sell their land or be expelled and move to the city. This period also marked the beginning of the country's industrialization – as the population left the rural areas and flooded the towns. With modernization, we see the birth of a working class and the emergence of an urban bourgeoisie. The evolving working class demanded social changes as it was growing. For the first time, we see the rise of political groups and syndicates outside of the two-party system: Liberal and conservative. Whereas the oligarchy was burgeoning the profits, it lacked the consciousness and the willingness to build a national economy. The Colombian oligarchy had a medieval, individualistic mentality regarding the economy, institutions, and market regulations. It hoarded profits and expanded for its gain. International exploitation of the Colombian working class became prevalent as the production of critical staple products increased.

The American and British investments in Colombia depended heavily on the land and the production of the export crops. The United Fruit Company, for example, settled in the Caribbean region of the country and employed many workers with a minimum wage and in conditions of semi-slavery and exploitation. The Colombian government understood the workers were exploited. The government was aware of the farmers' exploitation, but Colombians were paid less for their work than other countries in the region. The government had agreed to a lower wage than in the different areas of the Caribbean, such as Jamaica. The United Fruit had approximately 25,000 workers in 1928, and their pay was minimal on a working day of 12 hours. The people were weary of the poor conditions while the United Fruit Company profits. The political climate was unfordable for the workers. Interestingly enough, the clergy and the press were against the workers and tried to dismiss their dissatisfaction as influenced by the Russian propaganda brought from Moscow by some agents. [52]

On November 11, 1928, workers of United Fruit went on strike for one month, demanding better work conditions and pay. The strike gained solidarity in many regions of Colombia. The President of Colombia, Miguel Abadia Mendez, put General Carlos Cortes in charge of handling the strikers. On December 5, he fired on the workers who were sleeping on the square. It turned into a bloody massacre called "Masacre de los bananas."[53] Those who did not die instantly were finished off with the bayonet or buried alive in a mass grave. It was one of the deadliest and bloodiest responses to the fundamental right of protest, and the state was directly responsible for the massacre.

The next decade was followed by more violence. The land dispute had become one of the headaches for the administration of Lopez Pumarejo, whose approach to solving the issue was very progressive initially. However, the violence intensified as an offset of the Colombian landowners and the oligarchy to the agrarian reform (Law 200) passed by Alfonso López Pumarejo in 1936.[54] His presidency marks a series of agricultural reforms. Among others, Law 200 was meant to clarify land disputes. APEN openly opposed it since it encouraged the elimination of private property.[55]

However, Lopez Pumarejo modified Public Law 200 in his second term by giving the landowners concessions. The new "Law 100" was a setback in the previous agrarian reform. The land has been the center of many conflicts within or between states. The expropriation of lands dates to feudalism. It consists of the privatization of means of production by wealthy landowners to strip peasants from their property and turn them into wage laborers. In his work "The Capital," Marx discusses a concept called "primitive accumulation," to whom he refers to the takeover of small owners of arable land and their conversion into wage laborers. In this case, the Marxist framework warns of the birth of a class of expropriated peasants to wage workers or their conversion to the proletariat. The process, Marx says, was different in various countries, but the product of this revolution was the same. *The so-called primitive accumulation, therefore, is nothing else than the historical process of divorcing the producer from the means of production. It appears primitive because it forms the prehistoric capital stage and the corresponding production mode.*[56]

Until 1930, the conservatives maintained political domination over state institutions; nevertheless, the power distribution alleviated some tension. Thus, conservatives shared their decision-making power with liberals. [57] The year 1948 marks the violence's peak with the assassination of the former Liberal Party leader Jorge

Eliécer Gaitán. Gaitán was a complex figure, a Union leader who encouraged peasants to fight the Conservatives and the Liberals over land expropriation and "primitive accumulation."[58]. Gaitán was considered an enemy of both liberals and conservatives. [59] The protesting over the death of Gaitán spilled from Bogota to the entire country. It turned into intolerant social behavior, a culture, a collective madness that, once out of the bottle, was quite impossible to have under control.

This behavior was deep-seated in the mentality. It demeaned and belittled every aspect of human life as a right.

People were dehumanized, killed, mutilated, and punished as an example for others to see. Political intolerance created a morbid trend for years to come. The cost of human life was devastating. Over 200,000 people were killed during the years of violence. Both political parties involved their people in fighting against each other. Juan F Roman states: Extremely complex; the period was characterized by a heterogeneous mixture of electoral crashes and blind political rivalry, military actions, local vendettas, personal acts of revenge, economic extortion born of organized guerrillas,' and social banditry. However, the primary cause of this protracted period of internal disorder was the refusal of successive governments to accede to the people's demands for social, political, and economic changes. The conservatives employed private paramilitary to target liberals and poor farmworkers; the liberals and the Communist party used the guerrillas to fight the opponents. [60]

The 1940s until the 1960s marked a period of violent conflicts and social unrest in Colombia. Communities were devastated by bloodshed, killings, and irreconcilable differences between the parties. The conflict between the conservatives and the liberals had intensified as the ideological differences and wealth inequality caused by land expropriation fueled an armed and intolerant battle between peasants who supported each political party. In 1946, the Conservative Party won the elections by a small margin (or very peculiar circumstances due to a voted absentee). Mariano Ospina Perez, an elitist from Medellin Antioquia, assumed office. Ospina Perez represented the Colombian oligarchy and the wealthy landowners. Ospina Perez, one of the first founding members of The National Federation of Coffee Growers of Colombia, became its general director. His presidential agenda included the augmentation of coffee production and exportation. During his presidency, Colombia became one of the largest producers and exporters of coffee. His years of presidency were characterized by social unrest and a substantial clash with the Liberal elite.

The effect of the mishandling of the agrarian reform was devastating. It caused mass displacement of the indigenous population and farmers from their lands and homes. Unable to protect their livelihood from the ruling class, the indigenous population and the farmers turned to guerrillas for protection. The guerrilla numbers increased exponentially. The calls for a peace accord initiated by Rojas Pinillas[61] were somewhat successful. In the early 1950s, almost 7,000 guerrilla fighters surrendered their arms. The violence continued, especially after the paramilitary units started to engage.

Vast territories fell under the guerrilla revolutionaries and later were mined to keep the paramilitary and the state out. The corollary for the conflict was none other than the state itself at the service of the political class. It refused to face the problem of the displacement of the rural population and the botched agrarian reform. Within the proper definition of the state, people accept state violence in exchange for their security, but that was not the case in the Colombian conflict; instead, it was a double standard used by the state. During the years of the "Dirty War," state violence meant to protect the political elite and the (oligarchy) more so than protect the institutions and civilians. The institutions became a potent weapon in the hands of the political elite to protect its interests and crush any opposition where it could. The government's role in the conflict in the 80s and 90s caused more violence. The conflict escalated into an armed insurgency and a decimated territory under the control of the FARC, paramilitary troops, and narco-traffickers.

Discordant policies on land and property

The Legislative Act 1 of 1936 shifted the view on individual property. With the consolidation of power by the liberal party, the state adopted an interventionist policy on property. The Legislative Act 1 (1936) amended the Constitution of 1886, creating a legal framework for the state to decide on property, public, and individual use. López Pumarejo's Revolución en Marcha allowed the state to exert distributive justice and regulate the economy. It was legislated and constitutionalized.[62]

During his presidency, Luis Pumarejo focused on reimagining land and land distribution. To create a social net and public land, he emphasized creating lands for a social function. This reinventing of public policy gave Punarejo a legal and constitutional right to make a base for the upcoming agrarian reform. The inequitable land distribution since the country gained its independence from Spain and the consolidation of a wealthy rural class had deepened the gap between the minority landowners and most small farmers, mainly of indigenous background.

"Property as a social function permitted the expropriation of land not put into production by the owners; it would, on paper, be the foundation of the most ambitious agrarian reform in Colombian history."[63]

In the 1920s, the expulsion of residents and settlers from their land happened due to a need for arable land for coffee plant cultivation and the lack of law enforcement on property rights. Pumarejo tried to establish a precedent, a legislative and legal ground for the laws and policies on land. His centralizing policy meant to create state assets and a social safety net for the peasants and the indigenous population and prohibit further expulsion from their land.

However, Pumarejo's take on the agrarian reform caused an uproar among the landowners. It caused them to extend their borders by infringing on property rights laws. As a result, conflicts related to land intensified during the 1920s.[64]

Law 200 of 1936 -Colombia's first land reform- gave the peasants more advantages to increase their bargaining power over the land. The law also made it harder to expel the settlers from their land. However, the law did not foresee legal hurdles and procedures that would arise in the future. Also, the law did not provide provisions for the expelled farmers and laborers to claim the rights of their lost property.

Hence, the loophole in the law caused dissatisfaction among the peasants and small farmers. It did not address and solve land issues; furthermore, it marked the beginning of 'La Violencia", one of the bloodiest periods in Colombian history. [65] The Violence was a rural revolution and a deep divide between left-wing and right-wing ideologies represented in various population segments. In addition, it showed how the oligarchy and the Colombian political elite had driven the country to a total collapse – over the land distribution.

From 1953 to 1955, the conflict changed in nature. Those favoring the rural and small farmers became insurgents, as we see later in the 1960s. The insurgents were called Fuerzas Armadas Revolucionarias de Colombia - FARC. They remained in the rural areas and established a formal social order cooperating with the villagers who accepted their rule and ideology. As for the doctrine, the insurgents borrowed elements from Marx's theory on class warfare, class analysis, and primitive accumulation. As we will see in the following chapters, FARC was heavily influenced by the Soviet Union and inspired by the Cuban Revolution. Yet, its dynamics worked for Colombia and its represented cause.

The insurgents operated in the countryside. The disdain for the state's reluctance to appease the rural conflict and the anger over the lost land prompted many groups to form guerrilla units. In more than a dozen armed groups typically operating in the country, the most significant and capable fighting force was the FARC-EP. The insurgents deployed various tactics, including kidnappings, bombings, and military raids, to advance their goal of establishing a revolutionary government to redistribute wealth. As we will discuss in the next chapter, FARC survived for decades, and despite the state's determination to dismantle the movement, it grew exponentially. Its demise came from various reasons, including boosting the Colombian military with Washington's assistance. Over time, as some of the top leaders were killed or captured, FARC lost some of its ideological fervor and turned increasingly to drug trafficking and extortion to maintain operations, mainly in the countryside.[66]

The social unrest incited by the Colombian government tells us that the hypothesis of the failed state proved correct. In the 1980s, a series of institutional botches, including car bombing, blowing up state buildings, and sequestering judges and officials, culminated with the capture of the Justice Palace by the guerrillas of M19 in Bogota.

What makes the Colombian conflict complex is the fact that the state fought alongside paramilitaries to protect the interests of the elites, particularly those who held vast amounts of arable land and whose industry was profiting at the expense of the expropriated farmers and indigenous population.

The state entered the conflict not as a legal, political unit with a monopoly over the use of force and violence to protect the institutions but as the ruling elite's interests. The state in the Colombian conflict behaved as a private entity to justify violence, eliminate the opposition, and crush dissent. Still, most of all, it targeted the indigenous population as a stronghold and a bastion of the insurgency. The event epitomized institutional failure and the state's inability to control the conflict and solve the political crisis.

The Colombian Elite and its Role in the Civil War

The role of the Colombian elite in the ongoing civil war is ambiguous. When discussing the conflict, experts focus on the political actors, but little is said or elaborated about the role of the Colombian agrarian elite during the war. First, it is essential to define the Colombian agrarian elite. Then, highlight the economic interests, financial assets, legal and illegal activities, and state support.[67]

The vast literature on the Colombian conflict suggests that the promotors of violence were the drug cartels and the insurgency, leaving aside the elite's involvement in its collaboration with the paramilitary and the state. The agrarian elite in Colombia were made primarily of cattle ranchers.

> *It is evident, for example, that a sizeable and influential section of the Colombian elite had a direct stake in the war. Many cattle ranchers were much more than indirect beneficiaries of the massive displacement of people from their land*

during the war. Instead, they actively participated in the war, often leading paramilitary groups.[68]

Hence, behind the conflict and the fighting actors, the cattle ranchers played a critical role in shaping politics and the economy—prioritizing the local and individual economy instead of strengthening a national economy. Francisco Gutiérrez Sanín and Mauricio Barón mention*: "It is practically impossible, for example, to find paramilitary phenomena without the staunch and organized support of cattle ranchers.'*[69]

Theoretically, the roots of the conflict emphasize the same indicators that we have used to frame the case to a reasonable extent. First, the elite theory indicates that the change in legislation favoring the landowners enabled wealth accumulation in a very unconcerned manner for the peasants and indigenous population depending on the arable land. Not only did it set society back to a primitive form of gathering capital at the expense of the peasants, but it also showed a lack of institutionalism and technocracy—paramount for a state to run.

Secondly,

The state created a political precedent that we do not see in consolidated democracies but in fragile states: the founding of a revolutionary guerrilla to balance the power and a sponsored vigilante by the oligarchy that later converted into a proxy for the state. The roots of the conflicts are attributed to the dispute between the two political parties and the extremist factions in either one. Yet, the displaced farmers, peasants, and indigenous people were the ones to suffer the blow of the botched agrarian reforms.

Left-Wing Ideology and Warfare

A leftwing revolution is considered sound and relevant if its foundation is Marxism and tied to the "Internationale." Perhaps from the lack of information, the public regarded the Colombian guerrilla as savagery with raging violence and brutality. They embodied some of these traits. Still, their ideology, leadership, and strategy were left out of the discourse. Hence, at the beginning of this chapter, I introduce Karl Marx's theory to lay out a platform where I will build a narrative about the rise of Colombian Marxism, the attempts to create a "utopian" agrarian society influenced by the Chinese Agrarian Revolution, and land distribution. To understand how a left-wing insurgency operates, one has to realize that ideology fuels the insurgency and the populace affected by social injustice. The basis of every left-wing insurgency remains Marx's theory on class warfare and primitive accumulation and the idea that "the only means to solve the conflict are the revolutionary methods and the models offered in the early 20th century such as the October Revolution and The Chinese Agrarian Revolution."

In this book, I argue that the conflict was a halfhearted revolution within a more significant conflict. It was based on class dynamics, inside politics, international actors, and geographical features, though it never translated into a transformative process. Instead, the most prolonged insurgency expanded in number and scope but stalled in strategic planning.[70] Hence, it lacked a clear strategy to reenter the country's politics, appease class warfare, and become a serious opposition instead of a drug–financing insurgency.

Marx's Class Warfare and Primitive Accumulation

From a "dialectic" viewpoint, class struggle is an evolving conflict that will eventually lead to a revolution. The cause of this socio-economic development is dependent on how the bourgeoisie behaves. The bourgeoisie or the oligarchy—otherwise known as the ruling class, whereas national or international—will collapse due to modernization, the birth of proletarian consciousness, and the need to change the status quo. In simpler terms, the Marxist view on class stems from the dialectic principle that conflict is inevitable.

Regarding class analysis, Marx presents a normative system of what he believes to be valued. Still, they are moral assumptions that Marx considers the core of an egalitarian society. Class, to Marx, is a dialectic concept where conflict arises within, and it fuels the system with its conflicting nature. It is necessary -hence- to move society toward progress. According to the theory, people "dream" of a utopian, egalitarian society where goods, products, and services are evenly distributed. These assumptions are based on social norms but mostly on values and morals that a society should be built upon.

Class is an antagonist construct and ambiguous at times. Nevertheless, based on its antagonistic nature, class is a vital part of the capitalist system and cannot be conceptualized outside of it. According to Marx, it is paramount to understand what fuels the capitalist system. Dialectical materialism deals with how a system's components interact with one another. The output reflects the consequence of the class struggle inside the system. Marx bases his theory on a rigid, gear-shaped mechanical system that will eventually collapse if the distribution of goods cannot sustain a radically egalitarian society.[71] In other words, in his thesis about class, Marx assumes that society flourishes only when the material goods are distributed evenly between the members of the society regardless of their participation in the production process and their economic status. Thus, a classless society can only meet the needs of its members by providing equal access to goods and services. [72]

Marx argues that it is economically possible to organize a society where the radically egalitarian distribution of wealth can be sustainable. To achieve that stage, Marx proposed a revamp of the economic system. In his view, such distribution cannot be sustainable by reimagining the present system. Thus, the revolution is inevitable – his premise is to solve rising socio-economic problems, change systems, and redistribute wealth. In Marx's analysis, class is part of the process, yet he explains that it is excluded from the output. [73] Marx worries that the radically egalitarian principle cannot be applied if the class does not partake in the distribution of wealth. If the class cannot extend its role and participation from the process to the output, then Marx deems capitalism an immoral system of values and unfair to the working class.

Based on the premise that class has an ambiguous place and role in the system, observing what happens in and during the process is paramount. Hence, what is the relationship between class and production, and what is the connection between class and material goods? From an economic point of view, the link between class and product is clear, yet this relation is ambiguous from a social science stance. The relationship between class and product is vicarious, and human relations within the structure are primary. These relations consist of the rights and powers over production.[74] The arrangement can also be described in social-relational terms: The people who participate in production have different rights and powers over the use of the inputs and the results of their service.[75]

"Just like in Mao's Agrarian Revolution, everything builds up to class struggle, the ferocious ritual that allowed peasants to obtain their true liberation by publicly attacking landlords and other class enemies." [76]

Similarly, the class analysis was the premise of the Colombian Agrarian revolution. It highlighted significant points, and the theory produced a hypothesis applicable in an underdeveloped agrarian society with a local elite and replicable in similar settings.

The Revolutionary Armed Forces of Colombia FARC

This chapter describes FARC and other guerrilla groupings as insurgents, armed forces, and critical players in the long conflict. I focus on the revolutionary framework, which explains why FARC, as a political organization, failed to take over or be part of mainstream politics and power-sharing agreements. Not only that, but a revolution comprises some form of takeover using force. Yet, FARC and other revolutionary groupings failed to seize power as opponents of the elite and the state and carry out the claimed agrarian reform nationwide.[77] In addition, how do we frame the agrarian revolution inside a country with a working-class, some form of industry, and a multi-billion dollar agricultural conglomerate that controls the production of exports? Albeit, framing the role of the FARC is as challenging as its political and social posture. In the light of explaining why FARC played such a crucial role and how that perspective changed over the years, I look at the argument from a different standpoint: a dichotomy between a strong ideology and, strategically, a weak revolution. I argue that the guerrilla warfare in Colombia is congruent with Gramshi's and Lenin's claim that not all revolutionary circumstances will produce a revolutionary outcome unless a robust political party leads them to a coalition among classes. In addition, this sheds light on the puzzle I presented above: The lack of cohesiveness and cooperation between strata with the same or similar repertoire of contentions.[78]

An insurgency is a prolonged political-military activity aiming to control a country's territory and resources through non-statal military force and illegal political structures. Insurgent activities include guerrilla warfare, terrorism, and politics, such as propaganda recruitment. The insurgency operations are designed to weaken government control and legitimacy over the country and debilitate or reduce its presence in zones controlled by the insurgents. Most insurgencies desire to maintain a particular area by forming a shadow government.[79]

Insurgencies are various depending on their political agenda. The common denominator of an insurgency is challenging the state in non-conventional or guerrilla warfare to meet its political goals. An insurgency perpetuates only if its cause is perceived as just, whether reformative, liberationist, or stemming from ethnic and religious persecution. Insurgencies fall into four categories. Depending on the nature of the conflict, insurgencies can be political, military, traditional, and urban. The type of insurgency will also dictate its strategies, territory or zones of operations, and leadership.[80] Insurgencies rise due to injustice, dissatisfaction with the government or the state, the country's politics, and the need for political change. Insurgency usually aims to change the status quo. Insurgents represent a part of the population or perhaps the entire population of a country or territory if the insurgency is liberationist in nature. For the insurgency to succeed and thrive in making a political impact, it must have widespread support in the zones it operates or countrywide. Insurgencies function as organizations and must have clear objectives to achieve their political agenda. Insurgencies must base their "existence" on the politics of contentions and grievance and have a strategy or strategic plan. An insurgency can adopt the following objectives, but not be limited to – in order to achieve their goals:

1. Limit the government's ability and enhance insurgents' ability to provide public services.
2. Obtain the support or the neutrality of critical segments of the population.
3. Isolate the government from international diplomatic and material support and increase global support for the insurgency.
4. Destroy the self-confidence of government leaders or cadres, causing their resignation or withdrawal. [81]
5. Reduce and neutralize government coercive powers while strengthening insurgent coercive capabilities. [82]

When analyzing FARC in this framework, we can see that it lacked greatly in some areas but was dominant in others. For example, in the zones where it operated, the rebels scored high in using infrastructure and services for the population they controlled.

Since the state of Colombia had inherited a solid political elite, it had created international alliances and a solid relationship with the United States. Hence, this relationship was weakened during the Clinton administration. As I will explain later in the book, the United States showed reluctance to get involved in Colombia's domestic conflict with the insurgents. The war was a hardcore ideological struggle between the right-wing paramilitary and the Marxist guerrillas more than a domestic conflict.

FARC played complex politics with tactics of kidnappings, members of the Colombian political elite or oligarchy, and lastly, drug trafficking. It was, without question, a violent insurgency. However, the stigma surrounding the FARC leadership and their members outshined their ideological standpoint. As a result, the public overlooked why militants joined the arms in the fight against the state, the oligarchy, the wealthy landowners, and the elite. Throughout this chapter, I explain how FARC seized territories and claimed "sovereignty" over the occupied zonas guerrilleras.

The FARC operated in 242 areas, making up approximately 22 percent of Colombia's territory. Peter Calvert gives a few scenarios to explain why revolutions occur. According to Charles Tilly:

> "*A revolution is a process in which the political direction of a state becomes increasingly discredited in the eyes of either the population as a whole or certain key sections of it*" (Calvert 1970:4).[83]

For the sake of argument, while discussing guerrilla warfare in Colombia, we must consider a few factors. First, the ideology under which FARC mobilized farmworkers, rural population, youth, and indigenous people - recruited them and created one of the most prolonged insurgencies in modern history. Second, the PCC (Colombian Communist Party) spread ideological propaganda among the population, especially the most vulnerable, those stripped of all property and human dignity. Third, the heads of the FARC were members of the PCC and self-declared Marxists or communists. There are, of course, differences in the tone, but not in the doctrine between ELN and FARC - it is the same ideology. Although both organizations are Marxists in theory, they lean heavily on the state-controlled economy once they seize power, following the Cuban model.[84]

According to declassified CIA files, the 1960s marked a period where the communist movement consolidated throughout Colombia. The Cuban influence was heavy, and the communists realized that traditional politics were inadequate to come to power. So, instead, the revolution was the only means to take over the country. [85] In addition, in the social justice area, both FARC and ELN shared the same totalitarian ideology, justifying totalitarianism as beneficial and good. The presence or absence of the state in the areas occupied by the movement also indicates how successful the scope of the operations was. In the case of FARC–EP, the state was quite out and non-present in the region where FARC had the most popularity.

In his book “The FARC: The Longest Insurgency,” Garry Leech writes: “*To this day, despite the frequent claims made by mainstream analysts that Colombia is Latin America's oldest democracy, the government in Bogotá, has never effectively controlled all the national territories.*” [86]

Despite the long battle, the guerrilla warfare in Colombia, although violent and ideological, did not produce the desired revolutionary changes. Even then, with the peace agreement of 2017, the revolutionary changes were farfetched, and the schism between the Colombian State and the FARC has remained the same. The integration of former FARC guerrillas into the country's life is slow and with significant stigma, while on the political field, FARC's popularity has dropped to 1-2 percent.

The focus of the argument to explain the trend will remain Huntington's take on the revolution. As he describes it, a rebellion, insurgency, or war of independence must not be confused with a revolution.[87] Huntington (1968:264). Calvert - argues that for a revolution to occur, the government must be discredited, and a new political force must seize and legitimize power. The outcome of such a revolution is evident once the process has ended. The transfer of power, in this case, is merely the process in which the other set of polity replaces one polity. [88]

More than fifty years later, the FARC members – primarily poor farmers and indigenous people are yet to be integrated into the country's life and politics. The peace agreement terms were never upheld, and the agrarian reform never came to fruition. FARC fighters have practically no skills except for fighting in combat.

To understand the role of the FARC and other guerrilla groups operating in Colombia, we must consider that the control of vast territories by the insurgent forces happened in very particular circumstances but failed to produce significant revolutionary changes. Charles Tilly argues that there must be a schism between the two groups for a revolutionary change to occur. Such a split can occur at various levels of political circumstances. For example, in a potential "politics as usual" scenario, the cost of displacing one polity by another is low, and the schism or the split is not as sharp as in other occasions –which we will discuss later. Other revolutionary settings involve coups –which do not displace - many polities. Silent revolutions can displace many polities while being low in cost. Lastly, great revolutions are expensive, with irrevocable splits, and can displace significant polities. The Marxist argument made by Gramsci and Lenin is that unless there is a vanguard political party and a class coalition, the revolutionary situation will fail to produce a revolutionary outcome. FARC resulted from frustration with the agrarian reform and appealed to the lower strata – significantly displaced farmers. According to CIA special reports – it was considered banditry and not a threat to national security and stability, at least in the early sixties. [89]

The guerrilla warfare in Colombia was diverse and complex. It operated in the urban and countryside. It had many groups varying from militarized forces operating in the jungle to urban guerrillas, primarily students and intellectuals. Urban guerrillas carried out large-scale operations, at times well organized. Urban guerrillas notoriously carried out car bombings and kidnapping of politicians and personalities until it culminated with the seizure of the Palace of Justice.[90] The rise of the guerrilla movement in Colombia resulted from two major developments. The first was carrying out actions, from car bombing to kidnapping, until the seizure of the twentieth century. Secondly, it was conceived from political violence. This movement was meant to transform itself into revolutionary action. By creating a military force, the guerrillas aspired to exceed the dominance of the state, combat it, and eventually replace it. The guerrillas rose as an auto-defense military group to protect the farmers and their territory. In either case, the enemy was the Colombian state. Their ideal objective was to turn the war into a revolution with a set of ideological beliefs, take over the state, and eventually replace it with a new political regime.[91]

Historically, the guerrilla groupings emerged in stages. The first-generation military units were the EPL and ELN[92]. The second-generation guerrillas were the M19, Movimiento Autodefensa Obrera, El Partido Revolucionario de los Trabajadores, and other groups operating in urban or the Colombian jungle.

As we went over in chapter one, the Colombian landowners gained control over the agricultural land after the independence from Spain, forming a new ruling elite made of wealthy ranchers. In the mid-1900s, the elite was structured into two major political groups, The Liberals and the Conservatives, which dominated the politics in Colombia until the end of the twentieth century. The differences between these two ideological structures consisted mainly of principles: Trying to fit ideologically with similar political systems. The two political parties disagreed on separating the state, church, and centralized or decentralized government. Very often, the disagreement between the two parties consisted of unforgiving violence. It was prevalent, especially in the mid-20th century. The parties' interests always favored the ruling elite. [93]

In 1930, the Colombian Communist Party was founded under the auspices of the Comintern. The influence of the international left was present and noticeable. During the years of violence, the PCC [94] recruited many poor farmworkers from rural central Colombia in the provinces of Tolima and Cundinamarca. The PCC operated as a mobilizing political structure to unite the peasants, directly threatening the Liberal and Conservative elite during the conflict. Thus, the PCC was instrumental in organizing the self-defense movement (movimiento autodefensa). The leaders of the peasant movement were also members of the PCC. [95]

By 1950, the communist revolutionaries regulated the expropriation of the land and distributed it primarily among rural and poor farmers. In an analysis, the World Bank has laid out the ransacking model in countries where civil war or conflict has impacted resources and has used these resources as primary means of survival and a profitable economy. Such is the case of Angola with the diamond trade and Colombia with drug trafficking. Ransacking primary resources is one of the main variables in the analysis. The model suggests that the guerrilla failed in the military endeavor, yet it turned the trade into a profitable business. [96]

The FARC grew slowly but surely. Its strategy was strictly guerrilla warfare, ambushing state military units, raiding farms to seize land from wealthy landowners, and distributing in the zones they controlled. However, FARC sought the opportunity to capture hostages and military equipment and secure food and supplies.[97] It prioritized survival over seizing power and taking over the country. The harsh reality was that the Colombian Army's effort to eliminate the FARC became an outside, existential threat to the movement. As a result, FARC established codes and rules to prohibit failure from within. It enforced, among others, severe punishments for robbery, raping, and treachery or informing.[98]

In 1997, FARC made the list of terrorist organizations in the US. Until then, FARC was a threat to Colombia's internal stability. Hence, when FARC adopted drug trafficking as a profiting industry, it became e a stability threat to the region, including the US. The money generated from drug trafficking funded FARC and its operations. Before the drug trade, FARC generated income mainly from taxes and served as security units for the narco-traffickers. After realizing that the resources were relatively accessible and the infrastructure available, it seized the opportunity by directly engaging in the production and sales of cocaine. The drug trade was a direct funding of the agrarian revolution, yet FARC did not have a known strategy – except for occupying and operating in rural Colombia. [99]

Since its formation, FARC arose as an insurgency in response to the ongoing political conflict and failures of agrarian reforms. As we saw in the previous chapter, the failure of the liberal reforms to create a system of fair land distribution gave rise to the armed insurgency. In addition, the Colombian government responded to the grievances of those whose land was appropriated by wealthy landowners using force instead of reforming the land ownership. Hence, FARC had an ideology to justify the means and their actions, conforming with Marxist theory.

The longevity of FARC for over four decades depended on the shift of their operations to self-sponsor the war. FARC did not lean heavily on ideology to recruit and finance the war in the last decade. Instead, by the mid-nineties, it sought to cash in drug money, turning into notorious drug traffickers.

FARC accumulated substantial funding from the drug trade to sponsor the agricultural revolution and maintain thousands of guerrilla soldiers enlisted in the movement. It revealed a deep social wound based on the unfair distribution of land and the expulsion of indigenous people from their property. It exposed class warfare and a divided society ravaged by violence and rogue politics. It also showed how fragile the State was as an actor and its ineptitude to have sovereignty and jurisdiction over the national territory.

Like many left-wing insurgencies, FARC arose as a grievance-based insurgency with a clear ideology. Its objectives changed throughout the years depending on Colombia's political developments. FARC became the most prolonged insurgency, and the world outside Colombia was not aware that the movement's core was agrarian reform. Hence, the Colombian Government has done tremendous damage by discrediting the core of the problem instead of finding common ground. As I mentioned, FARC adopted weak strategies and engaged in combat mainly in rural zones, yet did not make military advances in the administrative cities, except for carrying out remote operations. Although it failed as a revolutionary movement, it remained the most prolonged left-wing insurgency of the 20th Century.

Colombian Paramilitarism

Bloodshed, violence, retaliation, and intolerance reigned supreme in Colombia during the civil war. A stable, functioning society was farfetched. People were born during the conflict and have lived most of their lives battling to survive. Indeed, the state never assumed responsibility for escalating the war - nor punishing those who committed the crimes. The argument of a hopeless culture of violence embedded in the people's mentality has sufficed for many scholars to make the case that the struggle was inevitable and fueled by the liberals and conservatives, the two major opposing political parties. Various Colombian scholars follow the same path, according to whom the culprit of the conflict is the disagreement between two irreconcilable ideologies. Surly, the blame falls on the Colombian state certainly. A fragile or failing state will always outsource its capacities to other actors to fight insurgencies. Whether by hiring mercenaries or paramilitaries, the state will seek help from domestic or international entities to alleviate the conflict. From the 1970s until early 2000, the Colombian state mobilized and sponsored militia units to support its military operations fighting the insurgencies.

During the 1980s and 1990s, the country underwent radical changes socially and politically. The crisis was tangible and evident in rural parts of Colombia where the insurgencies were operating. Despite being a stronghold of the elite, Bogota - the country's capital - mirrored the political struggle while the state scrambled to protect the institutions and manage the deadly conflict. The enduring conflict forced the state to seek a feasible solution from non-government organizations. The paramilitary was a right-wing formation made of various units throughout Colombia. It eliminated execution-style indigenous people, farmers, and their ideological opponents. AUC ambushed entire communities by fusillading men and women mercilessly. Created primarily as a private security service, the paramilitary degraded into a deadly force, widely known and notorious for violating human rights and committing massacres.

The Colombian government was well aware of the crimes - yet it cooperated in many operations. While the left-wing insurgencies proliferated, the AUC became pervasive. Its operations escalated from local guardianship to a violent paramilitary war, referred to as "La Guerra Sucia": The dirty war. The mobilizing of these parastatal military units was a green light to massacre without distinction opponents, people, or entire communities. It certainly was one of the darkest periods in Colombian history: state-sponsored violence.

The paramilitary executed thousands of people coldheartedly if they were suspected of having ties with the guerrillas. They had the upper hand in the conflict. By outsourcing the power to the private militia, the state looked the other way when their units committed violence. The notorious Colombian prison "El Modelo" in Bogota was one of the graveyards where the paramilitary dismembered, killed, and disposed of bodies in the sewer. In contrast, the guardians and the prison administrators looked the other way.[100] Its multifaceted operation scheme offered services to the Colombian Army and international and domestic businesses and cooperated with drug cartels and various criminal organizations. The paramilitary had multiple interests and partners. In September of 2001 – a little after the 9/11 attack, the AUC (Autodefensas Unidas de Colombia) entered the list of terrorist organizations operating outside of the United States:

Initially, the paramilitary became part of the Colombian state doctrine in regional and national security. As a result, militias and paramilitary are determining factors in national and international security. Although they operate as rogue actors, the states rely on their operations to combat insurgency, silence opponents, and maintain control over the territories.

This chapter covers the paramilitary activity in Colombia during the armed conflict and its culmination during the 1980s and 1990s. It will discuss the relationship between the Colombian state and the paramilitary group of AUC (Auto-Defensas de Colombia), its origin, doctrine, and recruitment strategy. It will also cover the scope and magnitude of their operations, including human rights violations, international government entities' involvement, and private entities with the Colombian paramilitary.

As violent non-state actors, militias and paramilitaries are organizations and function as such. They can be top-down structures sponsored and created by the state or bottom-up structures created by communities to offer alternative security. There is a distinction between sponsored militia and spontaneous militia in this case.[101] A sponsored militia receives training, funds, weaponry, and intelligence from the state when the state and the paramilitary have a common enemy, ideological or existential. A paramilitary organization differs from terrorist organizations, guerrilla insurgency, and other violent non-state actors. Paramilitary forces are generally created as local guardians to close political and social gaps within communities. [102]

However, militias or paramilitary groups do not always seek to close these social gaps; instead, they advance their objectives and agenda. The nature of paramilitary forces or militias is very delicate, and depending on their size, origin, and scope of operations, they can support constituencies. Yet, on the other hand, paramilitary groups engage in rogue action and indistinctively *en masse* killing.[103]

In a fragile or failing state- insurgent forces, guerrillas, and organized crime seek to directly challenge the state by overthrowing or controlling parts of the national territory. Paramilitary actors can cooperate symbiotically with the state or act as the above insurgency to overthrow the state. Thus, the relationship between the state and the militia can be fluid and might change over time depending on their political agenda.[104] States sponsor militias and paramilitary to cover operations that the national army might not cover. It delegates its powers to these entities in return for pacifying insurgencies or terrorist organizations. These groups are private citizens and are not part of the acting military. Hence, some militia members might also be part of the military. Yet, while adhering to the militia or paramilitary, these officials act entirely separately from the government or the state's military missions.

The National Security Doctrine and the Rise of the Colombian Paramilitary

When subversive forces became a security threat to the Colombian State and the elite power, strategizing and launching a national security doctrine became paramount. As the guerrillas gained control over the substantial national territory, the State adopted a distorted view of the Cold War narrative to align the country with other Western democracies. Hence, ideologically- the government had to espouse an anti-communist rhetoric. The State of Colombia had to survive to remain in power - at any cost. Crushing rebels was not feasible for a state struggling with many insurgencies and ongoing social unrest. The ideological war demanded that it invent a common enemy of the state and the people. The Colombian military worked behind the scenes, formulating an overarching ideology and a doctrine specific to Colombia. First, it had to identify the collective enemy, their zones of influence, and their modus operandi. More than a left ideological entity operating in the jungle, the guerrilla revolutionaries were a violent group that threatened the Colombian State and the political elite. Most notably, with the rise of Marxism in Latin America, the Eastern Block was exporting ideas and mobilizing insurgencies to counter-balance the United States' influence and recalibrate the balance of power. Thus, inventing a national security doctrine and forming an ad hoc strategic plan to combat the guerrillas, the State triangulated the resources to fit this ideological trope – State, Army, and Citizens against the internal enemy.[105] With this move and rhetoric, the state made the citizens part of the solution to the insurgency problem by inviting private citizens to the deadly conflict it failed to control.

The Colombian Army Generals and military officials in the late 1970s launched a public campaign to mobilize the entire population against the subversion and insurgencies they called the internal enemy. The indoctrination reached out to all branches of the government and set the stage for more human rights violations. The most absurd aspect of the indoctrination was the involvement of the martial courts in the trials of private citizens' alleged being involved in the conflict.

The Former President of The Supreme Court, Dr. Alfonso Reyes Echandia, affirmed years later that the trial of civils in the martial courts became an expression of the National Security Doctrine.[106] It opened the door for human rights violations and led to corruption, lawlessness, impunity, unaccountability, and retaliation. Launching the national security doctrine became an effort to institutionalize state retaliation by violating the principle of judicial independence. Trying civilians in martial courts undermined the court system's impartiality and institutional autonomy.[107]

The mass hysteria over the national security doctrine fueled and propelled by the generals resulted in more violent means of solving the conflict. It escalated the war to a full-blown military operation against the insurgency and its supporters instead of appeasing the fighting parties and seeking a more peaceful solution. Rafael Samudio Molina – General and Minister of Defense who in the 1970s founded various paramilitary groups, had requested in a letter to the President of Colombia in 1987 non-binding military structures to fight the insurgency. Non-binding military structures are none other than state-sponsored violent organizations with no accountability and not bound legally. Thus, there was no institution to monitor how the paramilitary would fill in the gap for what the Army could not accomplish - as a state-authorized force.

The paramilitary operations in Colombia dated well before the late 90s when the Castaño brothers emerged as a political force seeking legitimacy and competing for power and territory with the FARC. In the 1980s, different self-defense armed groups emerged in the Eastern Planes region (Llanos Orientales). These structures served as private security services and persecutors of the leftwing sectors. In the 1990s, these groups were the foundation of the paramilitary network of various blocs. [108]

The concept of the internal enemy was not limited to the insurgency and the subversion forces. It included all the marginalized layers of society, mainly the indigenous population, homeless people, prostitutes, homeless children, felons, etc. It was referred to at the time as "social euthanasia" or "social purification." [109] See the Spanish version.[110] This type of social engineering was impossible in Colombia and warned of a rise of the ultra-right ideology that led to human rights violations, more marginalization, and impunity for the crimes committed during the conflict. A social experiment of this nature had taken the conflict beyond ideological boundaries. It was no longer a struggle to keep the power but a justified method of extinction aimed at particular segments of the population and mainly those of indigenous origin or the political opponents of the power elite.

Hence, the adequate strategy to hold on to the power was to launch rambling and incoherent claims about the scope and magnitude of the insurgency. The Colombian elite, fearing losing control, scrambled to stop the spread of the conflict at all costs, especially in major cities such as Bogota, Cali, and Medellin. In addition, it sought to create an ideological and combative[111] block against the left. It declared an enemy not only the insurgency but also the marginalized strata. In other words, the indigenous population became the enemy – in rural Colombia.[112] Thus, the jingoists in charge of drafting the national security strategy launched a baseless political campaign claiming the "internal enemy" (the ideological left) had infiltrated all sectors of the country's social life. According to the claim, the guerrillas operated in specific areas but had cells and agents in every political, social, and civil structure. The allegations also included civil organizations as well as religious ones. [113]

The final means of the political campaign against the insurgency was to set the stage for the Colombian Army and the paramilitary to enter the conflict. The paramilitary doctrine *(Cited in Tras Los Pasos Perdidos de la Guerra Sucia)* is found in the Colombian Army doctrine, which states that; "If a conventional war becomes dangerous, the Army can use paramilitary (militia forces) to complete the objective successfully.[114] The Colombian government did not have sufficient human or financial resources to combat the insurgency. Capacity building capable security forces to fight the insurgents was also out of reach for the Colombian state - and the army. Not only that but the complex nature of the conflict and the use of paramilitary forces to combat trained and heavily armed guerrillas were challenging. Thus, the state had to leave many municipalities to the insurgents without contest.

Emergence of paramilitary groups

The paramilitary groups, which began as armed defense groups financed by staunchly conservative cattle ranchers and landowners in rural zones, emerged at the height of the drug trade and developed an intimate link to the drug economy. [115]The militia groups in the river Magdalena "proposed" the first model of paramilitaries in

Colombia. In some municipalities, local elites organized and funded paramilitary organizations to fight against the FARC and ELN, with covert assistance from military officers.[116] Therefore, the grouping was a significant concern for political science scholars and a topic of interest in Colombian politics. [117] The seventies and the eighties marked the birth of the first paramilitary groups: Autodefensas de Puerto Boyaca (APB) and the Autodefensas Campesinas de Magdalena y Medio, which later emerged with AUC, Autodefensas Unidas de Colombia and were part of the same grouping until its dissolution in 2003-2006. [118]

Vasquez also mentions that these dynamics demonstrated the conflict in the periphery and revealed the backwardness of the development of these regions. In other words, the rural areas were part of the conflict. Instead of progress, these areas only knew chaos, disarray, and violence until a particular social order was established under guerrilla rule to install a shadow government and some normalcy. The conflict, among others, introduced the birth of agricultural capitalism.[119]

Vasquez also mentions that these dynamics demonstrated not only the conflict in the periphery but also revealed the backwardness of the development of these regions. In other words, the rural areas have developed into violent conflict, chaos, and disarray without any known progress. The conflict, among others, introduced the birth of agricultural capitalism. [120] It was one of the harshest forms of capitalism - rooted in a feudalist mentality.

The first wave of violent groups in the 1970s came as the result of the political instability, social mobilization behind partisanship, the expansion of FARC outside of their traditional territories, the re-emergence of ELN EPL in the conflict as well as the increasing drug trafficking in the Colombian society (González, 2014). As we saw earlier in Puerto Boyaca, the initial paramilitary groups emerged under the command of the Perez brothers in the seventies. Ramon Isaza was also one of the pioneers to recruit the paramilitary units in the "Las Mercedes" region, later known as Puerto Triunfo in the Antioquia region. He named the paramilitary units the "Shotguns" (Los Escaperos) in 1977.[121]

In 1980, Chepe Barrera formed a self-defense group in La Dorada (Caldas). Lastly, the San Juan Bosco Laverde case is in the Santa Helena de Opón (Santander) district. The unit was called "Los Tiznados" and operated under the command of Isidro Carreño, who, together with other groups in the province of Chucurí, held back the guerrilla. Many of these units were co-opted by the military. [122]

Negotiations with the guerrillas during Belisario Betancur's presidency (1982-1986) led to the expansion of the paramilitary groups. This approach infuriated the militia, which, backed by the private sector, opposed all reforms and negotiations with the guerrillas for economic and political reasons.

The second wave of paramilitary activism became prevalent during the governments of Gaviria, Samper, and Pastrana. The attempt to legitimize the M19, ELN, and PRT to enter the political scene escalated the conflict and intensified the political violence.[123] The political crisis, decentralization, and the local government's autonomy gave rise to the paramilitaries and strengthened their relationship with the production and trade of cocaine.

Furthermore, the failure of the Caguan negotiation between the Government of Andre Pastrana and the FARC - produced more political violence and resulted in a territorial expansion in the zones where the guerrillas and paramilitary respectively operated.[124]

It is worth mentioning that the second phase of paramilitary activity is more strategic and ideological than the previous phase. In the late eighties, the paramilitary units in Colombia had taken the shape of a counterinsurgent organization that assumed political and territorial protection against the insurgency. Thus, we are looking at a much more coordinated and unified AUC in the late eighties and early nineties.[125]

Corporates and International Partners in the conflict

The accusations that the Organization for Human Rights has made toward multinational corporations in Colombia are substantial. The (OTP) Office of the Prosecutor of the International Crime Court has long investigated the involvement of Chiquita Brands International Inc.

Chiquita senior executives have been silent about their involvement with the paramilitary and its many crimes and forced displacement of farmers and the indigenous population.

Senior executives knew well about the crimes committed by the AUC and the many paramilitary blocks acting under the AUC. Nevertheless, they continued to use the paramilitary as their business partner, providing its services and payments to various units that operated under the umbrella of the AUC. These blocs included Bloque Norte, Bloque Elmer Cárdenas, and Bloque Bananero. The settlements continued even after the paramilitary was declared a terrorist organization by the U.S. in 2001. "These three blocs directly perpetrated atrocity crimes in Colombia's banana-growing regions in Antioquia and Chocó (specifically, in the Urabá region) and Magdalena.[126]

The declassified CIA documents reveal that the U.S. Department of Justice warned Chiquita to stop making payments to the AUC and that the payments were considered illegal.[127] Regardless, Chiquita executives continued to authorize payments until February 2004. The violation of the U.S. DOJ policy while doing business with an FTO (Foreign Terrorist Organization) revealed the transaction's nature. It provided further evidence that the payments were not mistakes but intentional contributions to the AUC. As a result, the Colombian landowners and the international corporations became the most prominent enterprises in the second half of the Twentieth Century. Such a corporation included Chiquita and its subsidiary, Bandex.

> *As a corporation, Chiquita pled guilty in U.S. federal court in March 2007 to engaging in illegal transactions with the AUC following an investigation by the U.S. Department of Justice for making payments to a designated FTO. At the sentencing hearing, the same U.S. prosecutor also stated at the hearing: "What makes [Chiquita's] conduct so morally repugnant is that the company went forward month after month, year after year, to pay the same terrorists. Chiquita was spending money to buy the bullets that killed innocent Colombians off those farms.*[128]

The need to provide security for the investments in Colombia was outsourced by private security companies, which were, in fact, paramilitary units. Thus, agribusiness became one of the biggest promoters of paramilitary activity. Since it expanded at the expense of the small farmers and rural population, the cattle industry, the fruit companies, and landowners sought necessary security units against the inflammatory activity. Apart from the economic interest of the big cattle and agricultural production, the security issue was a political endeavor to provide military assistance in the financial sector to deter the guerrillas.

The economic sector influenced politics incredibly, and in the communication sector, the interests became frequent points of divergence among the political parties. [129] The state's involvement with the paramilitary extended to other influential industries. The military seemed to use the national security doctrine to justify militia as an extension of the private sector. Thus, the paramilitary's presence was not seen as a separate entity but as part of the Colombian military on an organizational chart. Inherently, the use of paramilitary and military forces in Colombia can be argued as a "right-wing junta" in the Colombian government and governance. It was double jeopardy for democracy and the fragile state.

The dominance of the right-wing junta in politics deteriorated the lives of rural communities, farmers, and marginalized people under the guise of national security doctrine. Thus, the Colombian state had created various mechanisms to train civilians to participate in the conflict.

[130] Still, militia in Colombia is a complex case. It operated on at least three levels. First, the Colombian paramilitary was a state-sponsored terrorist organization involved in the great conflict ideologically and militarily. Since the FARC was the common enemy of the paramilitary and the state, it expanded violence in the zones where the guerrilla operated. The paramilitary gained notoriety, massacring villagers and farmers, allegedly FARC sympathizers. Secondly, the AUC headed by the Castaño brothers – was backed by the business and the wealthy landowners. Expanding arable land by displacing poor farmers and families became pervasive and continued to be the root of the conflict. Finally, the paramilitary held business dealing with the agribusiness and became associates of narco-traffickers facilitating and securing the cocaine routes. Thus, the paramilitary proliferated from a network of security companies to a serious business partner. They cooperated with the Colombian state, US fruit companies, wealthy landowners, narco-traffickers, and US state agencies in covert operations. In the late eighties and early nineties, the paramilitary had the upper hand in the conflict and a solid and strategic position.

The aforementioned paramilitary groups dominated part of the Llanos Orientales region. Their support from the Colombian state legitimized violence, and many of the crimes went unchecked and unpunished. The paramilitary engaged in massive human rights violations, murders, massacres, extrajudicial killings, and the disappearance of thousands of innocent civilians. Apart from the atrocities, the paramilitary forced illegal recruitments of the population. The way they operated is more or less similar throughout the entire region. However, some differences had to do mainly with the location, demographics, and geographical factors. The forms of action and the victimization patterns of these groups had standard features and different characteristics, depending on where they were located in Colombia. Geographic, economic, and demographic factors impacted their military operations. The lack of institutional presence affected the dynamics of the armed conflict and its actors. The impunity influenced how the paramilitaries operated, their relationship with the civilian population, and The unwillingness to de-escalate the contention - made the militia one of the most pernicious enemies of the Colombian people and democracy.

Thirdly, the Colombian paramilitary – was both a top-down state-sponsored group and a bottom-up organization that protected the so-called communities. The AUC offered protection and services to the drug lords and joined the cartels by sharing profits from the drug sales and traffic into the U.S.

Due to its complex nature, many state and international actors were involved with the militia forces. For example, in U.S. v Chiquita, Colombia's most prominent U.S. fruit conglomerate admits to paying the AUC (Autodefensas Unidas de Colombia) from 1997 – 2004. However, in September 2001, President G. W. Bush declared the AUC a foreign terrorist organization and a national security threat to the United States. Chiquita then settled a criminal complaint by the U.S. Government and agreed to pay a $25 million fine.

Paramilitary Covert operations: By definition, covert operations are military actions in which the identity of the parties acting must be concealed or denied. The national security doctrine pushed the population to stand against the internal enemy. The paramilitary went unchecked and unpunished for years. It left a scar on Colombia's history and people's memory. *(See CIA declassified files in the footnotes.)*[131]

In conclusion, the Colombian paramilitary was prompted by the rise of left-wing insurgencies.

However, what makes the paramilitary more problematic is the state-sponsored violence and its abuses. The state failed to assess the separation between the rebellion and the civil population. That led to more abuses not only by the paramilitary but also by the state.

The botched strategy to combat the insurgency deepened the divide among various population segments—the presence of the paramilitary marked one of the most challenging tasks in bringing about a peace process. The paramilitary was one of the primary opponents of peace talks with the guerrillas, which perpetually prolonged the conflict.

Cartels Against the Ruling Elite

With the proliferation of drug trafficking, it became paramount that the US government seeks a solution to get a grip and control the borders, ports, and airports. The drug cartels' routes to traffic drugs were mainly concentrated in the Caribbean region, including Cuba, the Bahamas, and Mexico.

In the last 25 years, the DEA has simultaneously battled the drug lords and warlords implicated with narco-trafficking – especially internationally. Hence, it is often restrained by foreign laws and is out of jurisdiction. Corruption and the refusal of local authorities to cooperate with the DEA had made the issue more complex. Historically, the hot spot of drug trafficking was Vietnam in the 70s and Latin America in the 80s and 90s. The heroin trade from Afghanistan later became prevalent, especially in the European markets. Hence, going after the lords instead of the systemic problem might be a temporary solution to the problem. It sure dismantles cartels but does not alleviate drug production and trafficking. In the early nineties, the DEA introduced the "kingpin strategy," a methodology to target the command-and-control elements of significant drug trafficking organizations.[132] Initially, the DEA targeted the drug cartels in Medellin and Cali, Colombia, mainly focused on the Cali Cartel. At the time, such a strategy proved productive for the dismantling and disabling of the Medellin Cartel and its kingpin, Pablo Escobar. However, as the plan evolved and resulted somewhat successfully, Colombia and Mexico implemented it as a strategy. Nowadays, the kingpin strategy and model are still in use; though modified to various countries' dynamics, they still use components that target the higher chain of command.

In principle, the kingpin strategy was a method that scrutinized the drug cartel and treated it as a business model or organization. It developed an org chart with a critical chain of command, activity, production, transportation, and infrastructure. The drug cartels differed significantly from other commercial organizations regarding structures and systems. They manufactured and sold a particular product, just like other enterprises in non-drug-related businesses. The DEA reasoned that tackling the drug activity would be non-productive and dangerous. Other law enforcement agencies, such as the police, handle these operations. Based on their analysis and especially the cartel's higher level of operations, it all came down to the highest chain of command. For their futile nature, the cartels, regardless of their organizational structure, are not immune to the threats and disruptions from the environment or other outside agents. If a disturbance in the organization's system is highly likely, or if the leadership is ousted, killed, or on the run, it would be a more successful operation than going after the activity.

It was decided that better success could be achieved by focusing on the people committing the crimes rather than the crimes themselves, especially given that many transnational criminal organizations are involved in multiple criminal activities worldwide.[133]

By the 1990s, the drug lords operated out of Colombia and distributed in North America and elsewhere tons of cocaine. They had accumulated great wealth and sought ways to launder money by investing in construction or pharmaceuticals - such as the Orejuela brothers. The Medellín and Cali cartels dominated the cocaine business; they controlled and micromanaged every operation chain, from production and packing to distributing.

The best practices passed on by the United States to the Colombians, in the form of the kingpin strategy, were essential to the dissolution of the cartels. The Colombian application of the strategy focused on the destruction of the cartels—not to stop drug exports but rather for self-preservation since the cartels intended to destroy the government. [134]

Hence, following the kingpin strategy, by the late nineties and early 2000s, the Medellin and Cali Cartel were dismantled. The strategy proved to be successful in Colombia. With assistance from the United States, it eliminated two of the most notorious drug cartels and their terrorist and illegal activities.

Trafficking cocaine was a profiting industry that benefitted not only the drug cartels but was also used as a primary source to finance the insurgency. The drug network in the late eighties was a monopoly of the Medellin and Cali Cartel. Yet, as a destination, it included a network of organizations and affiliations from Colombia, Cuba, the Bahamas, Mexico, and the United States. Not only did it cause a health crisis and social epidemic, but it also became a direct threat to national security if left unchecked. The main concern with the drug trafficking leaving the shores of Cuba was the cocaine load and possible missile and arms trafficking. The matter became a priority that the US Congress urged to screen various security agencies for possible conspiracy and corruption.

Hence, drug trafficking had become a Washington problem that the US government had to deal with on many levels. The issue the US faced in the eighties and nineties in Latin America was the distribution and proliferation of illegal substances in the American market. In addition, the assistance that Colombia needed to combat the drug cartels, the left-wing insurgencies, and organized crime was paramount.

The violation of human rights, which international agencies had already reported, raised concerns. Colombian and US courts would have constitutionally challenged a possible military intervention in Colombia. Thus, it was farfetched and not feasible. Such a move would have exacerbated the relationship between the two countries and have given rise to nationalist sentiments and rhetoric. Also, such intervention would have damaged the US image and its fight for democratic values internationally. Thus, the two countries agreed that collaboration would be a more acceptable and productive way to combat the drug traffic and deter the insurgencies from using it to finance the war. Amongst the political turmoil and violence on many fronts,[135] the Colombian state in the early 1990s declared Escobar and the Medellin Cartel the public enemy number one.[136]

The case was no longer about criminal activity and drug trafficking. The head of the Medellin Cartel scuffled the Colombian state openly. It targeted the judicial system by undermining its legitimacy and the legitimacy of the Republic of Colombia.[137] With the drug cartels entering the political scene, the conflict had entered new territory – it was the battle of the elite against the cartels. This book argues that the elite will always seek ways of self-preservation and holding on to power at any cost, even if the country is headed toward war and devastation. It also explains why the elite pursued solutions outside traditional Colombian politics to eliminate the adversaries. Escobar understood the political game early – yet he followed politics to advance his ambitions. The experience was bitter, and he learned the hard way that the elite was a closed, impenetrable circle of upper-class officials. They rejected him, and Escobar unveiled the weak foundations of the Colombian state, which was run down by the immediate interests of a centennial war between conflicting parties.

> *"We have watched shaken and horrified as the Government of Colombia courageously battles for its very life in a full-scale war against the entrenched armies of the drug kingpins. What has been perhaps most shocking of all is that in Colombia, the power of Government has been, at times, nearly eclipsed by the power of drug traffickers. In regular times, criminal activity operates at the margins in most societies. For Colombia, however, these are no longer regular times. The drug cartels have long since crossed over these margins, feeding the country's economy*

and gaining, in return, some support from the public.[138]

OPENING STATEMENT OF SENATOR NUNN

The US. Congress 1989

Escobar's disdain for the political class was greater than his ego. It turned him and his cartel into a terrorist organization searching to destroy the political elite – he called oligarchy. His calculated actions crossed the boundaries of the organized mob, scope, and magnitude. During his reign as the cocaine kingpin, Escobar ordered execution-style assassinations of his adversaries – many of them were political figures or officials in the judicial system. He eliminated judges, politicians, police, military officers, and state agents to destroy evidence and eliminate witnesses who could testify against his criminal activity. More importantly, Escobar was concerned with the elite's political game and involvement with the extradition law as a preemptive measure. Such a law was meant to deter not only Escobar from drug trafficking but also to discourage his cartel from waging war on the Colombian state.[139]

Never before the national police, the national army, and their paramilitary proxy were mobilized in a block to seek and execute him on the spot. In fact, "El Bloque Busqueda"[140] did not seem to have intentions of arresting Pablo Escobar and bringing him to court – whether in

Colombia or the U.S.[141] Escobar incarnated the underground world – yet, in fact, he was an existential threat to the state of Colombia. Medellin Cartel's biggest concern was blocking extradition law. [142] Made up of various clans and families, the cartel challenged constitutionality and questioned the legitimacy of a law that most Colombians were against – the extradition law. The general perception and public opinion thought the law was unfair. The dissenters argued that Colombians should be punished for the crimes committed in Colombia. At the same time, the proponents argued that a US sentence would deter the cartels from proliferating drug trafficking to fund insurgencies and criminal organizations.

One of the challenges faced by the Colombian elite in the late 80s and early 90s was getting the Colombian Congress to pass the "infamous" extradition law. Such law pressed that the Colombian drug traffickers be extradited and prosecuted in the US instead of Colombia. The extradition law became an imminent threat to the existence and operations of drug cartels. In retaliation to the state's effort to pass the law, terrorist acts, bombing, killings, kidnapping, and other violent crimes of this nature have boomed. The state faced another battle and was prepared to use all means, resources, political actors, and national and international sources to fight the cartels.

Interestingly enough, the Medellin Cartel was, by definition, a criminal organization related to drug trafficking, but in reality, they had a *political objective and agenda*.

Hence, the particularity of the cartel was the scope and magnitude of its operations. It attacked the state and government branches and influenced the judicial branch. Thus, more so than a drug cartel, its delimitations were significantly extended in areas not typical for such organizations. The Medellin Cartel also competed with the state and the Colombian elite by influencing the population in elections.

As I analyze in this chapter, the cartel had its scope, ideology, and political objectives. Thus, when a criminal organization that operates in the shadows of society or underground seeks political changes using violence, it can easily fit the profile of a terrorist organization. The Medellin Cartel used all the violent means that a terrorist organization would use to score politically, change the legislation, and influence the prosecution of drug lords.[143]

It executed high-profile politicians, journalists, judges, army colonels, and magistrates - apart from rivals and competition in the drug market. Moreover, the organization never spared the most dreadful retaliation against foes. Car bombing, kidnapping, killing, dismembering, executing, torturing, and claiming responsibility after terrorizing the entire country in the late eighties and early nineties. Based on the circumstances, the cartel shifted from a terrorist organization to a rogue insurgency toward the end.[144] It recruited street people and the needy to commit crimes either by suicide bombing or execution-style assassinations to stop the extradition law.[145]

Some researchers hesitate to solve the big puzzle or even consider answering the question: Why would a state - involve and outsource many structures to execute a criminal, ignoring that the country is a lab of malicious activities in all aspects? Why would the Colombian state prioritize Escobar and overlook the country ravaged by the guerrilla insurgency? The media and the entertainment industry have only covered the case of Pablo Escobar by shifting the focus away from his political involvement and agenda.

The overall perception of the narco-trafficking network in Colombia is what we all know: A criminal organization that trafficked illegal substances and flooded the United States with cocaine - through specific routes, by air, by boat, and by land. This perception is not equivocal; it is only a partial view of narco-trafficking as a critical factor in destabilizing Colombia's social-political situation. The Medellin Cartel was an enemy of the Colombian state because of its prolific drug trafficking and its determination to destroy the elite and the political class that once rejected Pablo Escobar from becoming a prominent member of Congress and, therefore, joining their political circle. The conflict culminated with increased violence, many terrorist acts, the assassination of conspicuous Colombian politicians, and the opposition to the extradition law.

"Colombia will extradite these criminals of humanity," said Belisario Betancur in the Cathedral – after Rodrigo Lara Bonilla's assassination increased efforts in operations against drug trafficking and drug lords. Yet the steps were not as productive.[146] As the scope of the conflict expanded, the state sought solutions by involving foreign political actors, especially in the national security sector. Along with covert operations, the United States had a significant part in the predicament with Plan Colombia.

The US involvement was diplomatic and military assistance to battle the growing insurgency and the proliferating drug production and traffic. However, the ineptitude of the Colombian state and its political elite to handle the increasing violence became a threat to the region. As a result, the US sought to adapt its foreign policy in Colombia to appease the violence and stop the conflict from spilling over to other Latin American countries dealing with drug trafficking, corruption, and insurgencies.

The drug cartels and guerrilla insurgencies were not the only problems Colombia faced. The rampant paramilitary violence had hampered and put regions and rural populations at a disadvantage. Yet, until 2001, when it was labeled a terrorist organization, the paramilitary continued to commit crimes side-by-side with the state. The ideological flag and the anticommunist rhetoric were nonetheless a sham effort to disguise the inherent conflict implications: the preservation of political power by the elite and its infinity. Both drug cartels and the insurgencies were sworn enemies of the state and the political class. Wrapped with the ideological flag against communism and organized crime, the Colombian state sought financial, military, and intelligence assistance from the United States.

On the other hand, the drug cartels declared the state and the national police as the enemy of the Colombian people. They engaged in an ideological battle to destroy the political class. The Medellin Cartel, the most notorious criminal organization, was finally dismantled and killed by the national police with the help of the DEA and the paramilitary cooperation units headed by the Castaño Brothers. However, eliminating Escobar and his associates did not alleviate the problem. It left a vacuum that was immediately replaced by other organized criminal organizations such as the PEPES (persecuted by Escobar, among others, the Castaño brothers), other drug Cartels, guerrillas, and paramilitary.

In this chapter, I also argue that drug cartels, apart from operating a network of criminal activities, were in opposition with the state of Colombia and sought the destruction of the Colombian political elite, including liberal and conservative state representatives and judges. Moreover, the ruling Colombian elite saw in the narco-trafficking a criminal organization of international magnitude and an enemy that wanted its demise. The drug cartels' main objectives were to control the state's decision-making and influence Colombian politics in favor of their illegal activities.

Without calculating how many homicides his assassins committed, the cartel contributed to the death toll and violence epidemic in Colombian society in the late 80s and '90s.

Escobar had all the money that one could have, and with that money, he wanted to buy his place in the traditional elite and somehow make his connection with the crime and tag as a bandit disappear.[147] Instead, Escobar was rejected by the conventional Medellin elite and seen as a danger. Hence, Escobar could not launder money in any industry, even though he tried to buy a few factories. The more suitable sector where narcos were able to launder the funds was construction. Thus, the money poured into construction destabilized the economy, and inflation increased. [148]

A criminal organization is an entity that engages in illegal activity for profit. Such organizations have members, clear objectives, and goals. The structural chart of a criminal organization is more or less a top-down scheme. It has a unit of command or a person at the top of the hierarchy, making most decisions. Almost and with certainty, the decisions are made a priori by the head of the organization, sometimes consulting with other top members. The organization's roles are well defined, and every individual has an assignment for which they are accountable, an essential link in the entire chain. Although criminal organizations have their scope and magnitude, the authority figure has limited powers outside the organization. Hence, for the structure to be successful, it must have clear goals and objectives and adopt a system that will function regardless of interference from the outside and lapses from the inside. Criminal organizations are not entirely closed systems. For systems to function and live, they must deal with entities outside the organization. Yet, in criminal organizations, the areas of operation are illicit activities, corrupt individuals, and other organizations - licit or illicit.

Criminal organizations operate in the shadows of society and avoid contact with government agencies, police, prosecutors, detectives, and agents. The modus operandi of such entities relies on the selling and trafficking of illegal substances – drugs or stolen goods, illegal services, evading dues and taxes, fiscal evading, and illegal trade deals. A criminal organization also launders money regularly by investing in businesses not very tangible by the state and local government or agencies. Such businesses include investments in construction, shell companies, and overseas companies that do not declare their revenues or profit. Cartels are criminal organizations that operate by producing drugs and trafficking them in the market where there is demand. Defining a criminal organization is not an easy task, and when the organization operates outside of the United States- it becomes much more complex. Many variables come into play. In the case of Colombia - the drug cartels in the 1980s were fraught with many issues. They evolved as crime organizations and enemies of Colombia's people and the state.

Hence, the Medellin Cartel - one of the most violent criminal organizations, fell into various definitions. The cartel produced and sold drugs in the US and Europe. Still, it engaged in financial transactions with other criminal organizations. For example, the cartel members, such as the Ochoa brothers, had capital invested in the cattle industry, the emerald trade, art, and high-profile business trading among people of business. Thus, Medellin Cartel was well suited to be a syndicate with substantial financial interests in the Colombian informal economy. Street gangs ran the cartel's lower ranks with close ties and direct orders from Escobar. Regardless of his highest role in the Medellin Cartel, the three Rs of a street gang (reputation- and retaliation) were also the moving force and its modus operandi. Loyalty was a must, and retaliation was tenfold more remarkable than disloyalty. Since the 1980s, the US demand for cocaine, heroin, and their sub-products increased exponentially. Although there have been fluctuations in the market, the supply seemed to keep up with the demand, whether from Colombia or Mexico. The primitive thought and rationale behind the drug lords were prompted by the need to produce and sell drugs. In the late eighties, Colombia was the largest cocaine producer and exporter in the US and the European market. The war on drugs frenzy in the US did not seem to slow down production or traffic. It only changed the method and the routes. After Escobar's death, the heads of Cali Cartel and Norte Del Valle were deported to the US. Dismantling the cartels changed the game and cocaine trafficking forever. It indeed prompted international cooperation between criminal organizations. Drug routes, the magnitude and the scope of operations, and the influence in the economic sector shifted back and forth from Colombia to Mexico. Yet cocaine was still reaching the US streets and markets. The proliferation of drugs originating in Colombia did not change drastically, but the structure of drug cartels became less complex and more tangible. The drug production, packing, and transporting business strengthen the link between Colombian and Mexican cartels. From the 1980s to 2010s, the US became the primary market of cocaine coming from Colombia and Mexico.

It is paramount to understand what this chapter is not. Escobar's parable has been discussed, and many books have been written about his wealth, lavish lifestyle, vintage cars, painting, the stash of cash, etc. Yet, very few see Escobar as one of the political actors during the conflict.

Escobar shifted the Colombian state's priorities and grabbed international attention. This chapter covers his political involvement with the state and interactions with other political actors.

First, the chapter discusses Escobar's role in escalating the conflict, meaning it will analyze his involvement with the guerrillas and their joint violent acts on state institutions. Secondly, it will cover the relationship between Escobar and the Colombian elite, mainly his disdain for the neoliberal political class.

This chapter also discusses the open conflict between Escobar and the Colombian Press as the extension of the liberal wing in Colombian politics.

Fourthly, it will discuss the mobilizing of state agencies, military, and paramilitary to assassinate Escobar and destroy the Medellin drug cartel.

It will discuss his opposition to the extradition law and the US involvement in Colombia's domestic affairs. Finally, it will cover the DEA activity in Colombia to limit drug trafficking.

The Rise of the Medellin Cartel

The 1980s marked the rise of the international drug trade. The drug lords and kingpins had already established a market, infrastructure, raw materials, and an entire industry for the production and distribution of cocaine.

"Drug trafficking organizations reorganized and began operating on an unprecedented scale. The rise of the Medellin cartel, the influx of cocaine into the U.S., and the violence associated with drug trafficking and drug use complicate the task of law enforcement at all levels. Violent crime rates rose dramatically during this period and continued to grow until the early 1990s."[149]

In the 1980s, cocaine entered the drug market as the drug choice of the affluent and the middle class, and it was spreading rapidly. It became the choice among the upwardly mobile population, bringing drug use to unprecedented and dangerous levels. [150]

In the early 1980s, Escobar was well on his way to becoming the most prominent drug lord for the next decade, reigning supreme in Colombia and dominating the global drug market. The Medellin Cartel was already established as a criminal syndicate. They used murder, intimidation, and assassination to keep journalists and public officials from speaking out against them. By 1985, Colombia had the highest murder rate in the world. [151] The DEA reported:

In Medellin alone, 1,698 people were murdered, and the following year, that number more than doubled to 3,500. The Medellin cartel was fast becoming the wealthiest and most feared underworld crime syndicate the world had ever encountered. During the period 1980 through 1985, the Medellin mafias delivered to the U.S. a considerable measure of the wholesale violence and terror that their drug trafficking activities had inflicted on their home country. In a grim parody of their campaign to control Colombia, they insinuated themselves into legitimate and valuable sectors of the U.S. economy, such as the banking and import industries. The U.S. suffered severely from the cartel's presence as local drug gangs began to form, communities were terrorized, and drug use among teens continued to climb.[152]

The Cartels against the Colombian institutions

The siege of Colombia's courthouse (Palacio de la Justicia) on the 6th and 7th of November 1985 marks the culmination of the institutional crisis in Colombia. The conflict transferred from rural areas to the metropolis of Bogota. The siege put a question mark on the future of Colombia and the political class. The reality seemed gloomy; the carnage, hostages, and bloodshed inside the palace proved the Colombian state's fragility and unpreparedness to handle the crisis. The country was perhaps hours from collapsing had the guerrillas coordinated various attacks on other institutional buildings.

It was a turning point for the Colombian political class. Whether they take back the Court House and release the magistrates, or the state would fall under the rule of the insurgents in perhaps a matter of days. Even though the attack did not intend a government takeover by the guerrillas in some form of a coup d'etat, it weakened its position, destroyed its reputation, and deepened the conflict. It was a slap in the face for the Colombian Military, who had worked for years behind the scenes to mobilize all Colombians against the leftwing revolutionary insurgents. The event almost brought the Colombian state into chaos. It targeted the highest institution of the judicial branch and caught the entire political elite off guard.

There are, of course, speculations about the planning of the siege. Various political groups blame the attack on the Medellin Cartel and M19 former leader Carlos Pizarro. [153] Yet, Escobar never assumed responsibility, nor did the M19. On the contrary, the guerrillas claimed that the association with the drug cartel hurt their purely political cause. Various sources close to Escobar, such as one of his assassins, Jon Jairo Velasquez, claimed Escobar paid the guerrillas about 2 million dollars before the attack. In his 2002 book “My Confession,” paramilitary leader Carlos Castaño made the first claim that M-19 leader Carlos Pizarro received 2 million US dollars from Escobar in May 1985 to carry out the attack later that same year. [154] The attack was carried out by M19 guerillas only.

The inimical relationship between Escobar and the Colombian elite

Alonso Salazar. J, a Colombian politician (former Mayor of Medellin) and at the same time the author of the book "Pablo Escobar, El Patron del Mal" reveals that Escobar developed a complex relationship with the Colombian political elite before and after his debut in Congress as a Congressmen representing Medellin and the Liberal Party in 1982.

The presidential race between Belisario Betancur (of the Conservative Party) and Julio Cesar Turbay of the Liberal Party revealed that the money from narco-trafficking had already entered politics. At the time, Salazar reveals that Colombia was just a developing country, a backyard of the United States, tired and exhausted from the political war, the hegemony of two political parties, and an ongoing insurgency. Nevertheless, Turbay won the electoral race, and it was a backlash for the conservatives who basted Turbay for his connections with the narcotics traffic.

Before the violence and the war in 1958, the two parties signed a power-sharing agreement to rotate their governing every four years. Yet in 1970, a non-traditional left-wing candidate, Gustavo Rojas Pinilla (ANOPO), the National Popular Alliance, and Misael Pastrana, - the conservative candidate, entered the race. After Pastrana won the election, a revolutionary faction of the ANOPO group decided to form the M19 guerrillas out of outrage for the injustice – assuming that the election was a fraud and stolen from the left-wing ANOPO. The left-wing revolutionary group held the name of April 19. M19 became one of the most prominent guerrilla groups that were not only very eloquent in the political discourse by its leaders, such as Jaime Bateman, but they carried out substantial operations following attacks, kidnappings, and executions. Escobar found himself attracted by their whimsical revolutionary rhetoric and their boisterous actions. More importantly, Escobar found something in common with the M19 guerrillas. They were both sworn enemies of the Colombian state. At the same time, they were targets of the United States foreign policy in Latin America and the war on drugs – promulgated by the Nixon administration and later by Reagan.

In the late 1970s, political unrest in Colombia continued as outliers left the traditional political parties and sought alternative solutions. Such was the case of Luis Carlos Galan, a former education minister and Congressional Senator who, together with Alberto Villamizar, Rodrigo Lara Bonilla, and Enrique Parejo, formed the Neo-Liberal movement in Bogota. Galan's vision included war on clientelism, campaign financing with drug money, impunity, and justice for the Colombian people.[155]

Escobar's aspirations for a political career ended when he decided to assassinate the minister of the justice department, Rodrigo Lara Bonilla. Ironically, Escobar had all the money he needed for a political career, yet it was not enough. In fact, with the money and the political career, Escobar wanted to buy social status and erase his humble roots and his past as a bandit. Yet, the traditional elite rejected him and saw Escobar as a danger. Instead of opening the doors for him as a junior representative in Congress, they denied him access to many political circles, including Club Campestre de Medellin. Seemingly, there were no enterprise opportunities and connections between the new class of drug dealers and the traditional elite despite the ability to invest.

The Colombian elite was a closed circle and very conservative in its choices, yet it was a social stratum built since the times of Bolivar. The new drug kingpins were still considered ordinary people or low-class bandits. Thus, with tremendous capital but no opportunities to invest in industries, the Cartels invested in construction and buildings. Informality was rampant, and it spiraled out-of-control inflation.

After the scandal with Rodrigo Lara Bonilla, the government started to scrutinize the involvement of drug money in politics. The scandal had no precedent in Colombian Congress and left a bad taste in traditional politics. After Lara Bonilla discredited the allegations made on his behalf, Escobar was expelled from Congress, and his US visa was revoked.

Rodrigo Lara Bonilla had lifelong experience investigating criminal activity in Colombia and the surrounding countries. The struggle with Escobar crossed the boundaries of politics, and their fight became more character assassination and tactics to silence the opponent. Bonilla went full-fledged on the Medellin Cartel, ordering the seizure of planes' properties used to distribute illegal substances. Soon after, Lara Bonilla filed charges against Escobar and Carlos Lehder. [156] Thus, the traditional Colombian elite declared war against the narcos. The only answer that Escobar sought to solve the problem was the assassination of Lara Bonilla. The nearsighted solution began one of the darkest periods in the Colombian conflict.

The war between the Medellin Cartel and the Colombian state intensified until the last one was destroyed.

It was as bloody and violent as the other wars insurgencies were fighting simultaneously. Many victims of violence, bombing, and explosions are yet to be found, or their families to come to terms with the pain and suffering that the war caused. The brawl between the narcos and the state pushed the passing of the extradition law without questioning its constitutionality. The Colombian elite could not control the many open fronts to hold on to the power. The rampant impunity, corruption, and lawlessness were at an incredible point. It was, in fact, the result of years of violence and elite domination for power without governing and using the state in its favor instead of protecting the institutions and the citizens.

Terrorist attacks on the people of Colombia and State officials[157]

Struggles to capture and prosecute the drug lords in the court of law became fraught with complexities. Many lived in the outskirts of Colombia with high-security systems and were always on the move. The impunity, the corruption, and the dysfunctional judicial system had given the upper hand to the Medellin Cartel and its members to prolong the fight against the extradition law. The Colombian state was left with no other option but to draft a law that would rid Colombians of drug dealers once prosecuted in the US. However, the law was debated. Its constitutionality was questioned since the public did not entirely agree that native Colombians be tried in an American court by an American jury and serve their time in United States prisons. In retaliation for the extradition law, in the early 1990s, the Medellin Cartel waged war on the state to pressure the Colombian government to prohibit the extradition of native Colombians to the US. [158]

Headed by Pablo Escobar and several other members self-labeled as “The Extraditables,” the cartel took increasingly violent action to pressure the government to strike down the extradition law and draft legislation that would protect them from extradition. Among blowing up institution buildings, cars, and commercial fights[159], the cartel was responsible for the assassinations of dozens of government employees and the corruption of many state agents, among other high functionaries and elected officials. Henceforth, in July 1991, after many acts of violence and terror, the Colombian Congress gave up the fight and amended a constitutional law to prohibit the extradition of the Colombian natives. The move was a victory and highly favored the drug cartels since the extradition law aimed to strike the drug cartels.

Regardless of the short-term victory, the cartel started to decline. The Ochoa brothers – critical players in the Medellin Cartel – surrendered to the police and struck a deal with the prosecution to serve a lesser sentence. Escobar's assassins and other members were either killed or ran from the National Police and other adversaries. However, the many law enforcement efforts to topple the Medellin Cartel resulted in numerous surrenders and arrests that eventually led to the cartel's demise.[160] The Prisco brothers[161] – Escobar's terror squad were the most wanted for ordering the murders of 50 Medellin police officers, for several terrorist bombings, and for nine assassinations, including that of a Colombian Justice Minister in 1984.[162] Hence, in June 1991, Escobar surrendered to authorities and was sentenced to Envigado prison.

In Escobar's surrender negotiations, the Colombian government had agreed that security at Envigado prison would be the responsibility of Army guards and Escobar's handpicked bodyguards. In reality, Envigado prison protected him rather than incarcerated him. Escobar's period of imprisonment was considered his "Golden Age," during which he ran his drug empire without fear of being hunted by the Colombian government or assassinated by his rivals. [163]

In July 1992, Escobar "escaped" from Envigado prison to avoid being transferred to a Bogota jail after it was confirmed that Escobar had ordered the murder of some 22 of his drug mafia associates. One or two of Escobar's victims were even tortured, killed, and buried on the grounds of Envigado prison.[164]

Escobar had a warning of the plan to transfer him to a more secure prison, and 28 guards were later charged with aiding and abetting Escobar's "break out." For 17 months, Escobar was the target of the most extensive search in Colombian history. Finally, in December 1993, the CNP killed Escobar in a firefight at a private residence in downtown Medellin. Escobar's death and the surrender and arrest of the Ochoa brothers marked the decline of the Medellin cartel.[165]

Militarizing the State

The United States' role in the Colombian conflict was paramount to alleviate the pressure and to assist the state and the military in defeating the insurgents and the cartels. The puzzle that must be solved or examined involves political events shaping Colombia's domestic and foreign policy. The United States played an essential role in boosting the Colombian military. The perception that the Colombian political and traditional elites used US assistance to preserve their existence was not far from the truth. Colombia received more military aid than any country in Latin America. [166]

A historical perspective between the two countries reveals that the US assistance to Colombia's military began in the 1940s. The United States trained Colombian Army officers in the US military schools, boosting their capacity, especially with the rise of the insurgency. The long-time cooperation consisted in many areas, and the US support poured onto Colombia's national security to invigorate the fight against the narcotics and the left-wing rebels. The military agreements between the US and Colombia started right after WWII, and it soared in the early 1950s with the deployment of a Colombian combat battalion to the Korean War.[167] During the Korean War, Colombian soldiers fought along with the US units. [168]

The "interfering" in Colombia's domestic affairs was not a unilateral decision nor imposed by the United States. Instead, it was a politically calculated move by the Colombian political elite, especially by right-wing politicians. The US assistance has been regarded as a proxy to help combat both ideological insurgencies and narco-trafficking.

In 1982, the Reagan administration concluded that the Medellin Cartel, headed by Escobar, was a threat to U.S. national security. The use of narcotics in the 1980s became rampant. It incentivized the Colombian drug cartels to increase production and trafficking. By the 1980s, coca production was handled in Colombia. The drug traffickers no longer needed to get the coca paste out of Ecuador. Colombia also had the perfect climate to grow coca crops. The business became very profitable, even for small farmers. It did not require long cultivation periods, though it remained risky.

The drug problem was an existential threat to the Colombian government. The cartels became challenging adversaries. They were too strong to handle by local and national law enforcement agencies. Their network consisted of thousands of people, from dealers, traffickers, distributors, and security to assassins. The crimes committed by Escobar's assassins and his "death squads" dominated the political and crime scenes in the 1990s. Ironically, their operations were successful and often ahead of law enforcement agencies. The crimes committed in collaboration with the insurgents or the paramilitary were funded with drug money.

> *In 1986, President Ronald Reagan passed Presidential National Security Directive 221, which declared illicit drugs a threat to U.S. national security and gave the U.S. Department of Defense (DOD) a counternarcotic role of unprecedented size. The administration of George H. W. Bush further increased the military's role in South America by enacting the National 8 JSOU Report 14-3 Drug Control Strategy of 1989, which allocated over $1 billion in assistance to military and law enforcement organizations in the Andean region.*[169]

The funding went to train Colombia's military and police for operations against the cartels. It was a comprehensive packet of special operation training by US Navy Seals, US Marines, and the DEA that provided trading for the military, national police, and special agents. The training included jungle operation skills, surviving skills, operating combat, and weapon systems. [170]

1992, the Colombian government created "Bloque Busqueda," or the Search Block. It was a Joint Special Command created to fight narco-terrorism and narco-trafficking. Bloque Busqueda proved to be successful in its operation against the drug cartels. It dismantled the major criminal organizations in cooperation with the U.S. Special Forces in almost two years. Thus, the drug revenues were no longer the monopoly of the major drug cartels, which allowed FARC to enter the drug industry and bargain for better deals.

By the mid-1990s, the U.S. and the Colombian military had increased their cooperation in combating narco-trafficking. It was part of a multinational operation. The American Special Forces continued to boost the Colombian military by providing technology and devices used in air traffic. In addition, the U.S. Air Force radar sites assisted in aircraft tracking and shared information with the Colombian military. The American Special Forces helped Colombian police and the DEA intercept illegal activity, especially the drug laboratories in the jungle. On the other hand, the FARC and the ELN had expanded their control in the periphery and rural areas. The insurgency used Maoist tactics and established shadow governments operating parallel with the Colombian government inside the national territory.

The insurgents took over hospitals, schools, part of the infrastructure, and roads in these areas called "zona guerrillas." Their operation increased. The Colombian government was in a dire and challenging position. The collapse of the government was close to becoming a reality. They sought to use the help and assistance from the U.S. to boost the counterinsurgency efforts. Still, the Clinton Administration did not share the same view on ideological wars in Colombia. The United States had offered funds and resources to combat the narcotrafficking, yet the insurgencies were a domestic problem.

Moreover, the Clinton administration was concerned mainly with the drugs that were leaving Colombia for the United States, not guerrillas whose ideology had been dealt a mortal blow with the fall of the Soviet Union. [171]

As a result, the U.S. Government continued to support only the Colombian National Police. Consequently, the United States concentrated its aid on the Colombian National Police and insisted it be used for counternarcotics purposes. By 1998, 90 percent of U.S. counternarcotics assistance went to the Colombian National Police.

The need of the Colombian Elites to turn Colombia into a protectorate of the US was an effort to boost up the military with direct assistance and financial aid to combat the insurgency and narco-trafficking. In the late nineties, Colombia was in a dire political situation, and the relationship with the United States had soured due to the human rights violation reports and the ineptitude of President Samper in combating the insurgencies. As a result, Plan Colombia offered many incentives for the Colombian Government to regroup and gain self–confidence.

> *But in critical ways, the U.S. intervention tipped the war. It delivered a shot of confidence to Colombia's institutions, particularly its military. It gave the country a vast, sophisticated intelligence-gathering system to hunt the rebels and the lethal hardware to strike them from the skies.*[172]

By 2003, nearly 5,000 staff members and private contractors were working out of the American diplomatic compound in Bogota, making it the largest US embassy in the world.

Hence, the critics of Plan Colombia call the "US boosted strategic plan" to combat the narco-trafficking and the insurgencies – an interventionist move that in some way helped only one side of the conflict – that of the Colombian political elite, the international corporates, and the right wings groups backed by the Colombian Army.

Unfortunately, it boosted the paramilitary and gave them the green light to massacre civilians alleged to sympathize with the FARC. Washington Post writes:

> *With U.S. backing, the Colombian government launched a scorched-earth counteroffensive against the FARC's rural strongholds after President Álvaro Uribe was elected in 2002. Government troops were often followed by right-wing militias that targeted suspected rebel sympathizers and massacred civilians. More Colombians were driven from their homes during the first stages of Plan Colombia than at any other time in the half-century conflict.*[173]

The political class's ineptitude to govern and control its national territory brought in measures that prompted the US to use interventionist politics. It facilitated and allowed a foreign country to combat terrorism and ideological insurgencies instead of its military and armed forces. Even though there are no official reports from the Pentagon of American troops fighting in Colombia, many military training and strategies took place during Plan Colombia instead. In addition, testimonies of human rights violations revealed that Colombia was the number one country for crimes that were unsolved due to ineptitude and corruption.

Alas, various agencies' involvement with the covert operations was never a secret, whereby the United States government backed the Colombian government in the fights against the ELN and FARC. The covert operations indeed deterred the FARC's activity by targeting their leadership. Hence, thanks to the US assistance, the Colombian military was able to eliminate more than two dozen rebel leaders. [174] The dividing line between narco-trafficking and insurgencies was challenging to define since they intermingled and had a similar modus operandi.

In 2000, the National Security Agency launched a surveillance program in Colombia aimed at deterring the FARC and its operations. Therefore, nine billion dollars went to Colombia through "Plan Colombia." Much of these funds were used for military aid and assistance. Among the Congress-approved funding, there were reports of undisclosed funds for secret operations approved by the George W. Bush Administration. With the growing FARC presence, especially in rural Colombia, and their bold action against political candidates, the U.S. feared that Colombia would collapse.

By 2003, U.S. involvement in Colombia had increased exponentially. It comprised 40 U.S. agencies and approximately 4,500 people, many of whom operated out of the U.S. embassy in Bogota. It was the country with the most extensive military aid assistance for at least two decades - besides Afghanistan. The U.S. ambassador to

Colombia from 2003 to 2007, William Wood, said, "There is no country, including Afghanistan, where we had more going on."[175]

The covert operations in Colombia continued well from 2001-2007. With the new technology, the Colombian military, guided by the US and Special Operation Forces, could locate the FARC activity with precision. Yet, capturing the rebels was harder. Many times, these operations proved to be unsuccessful. While the detection of FARC training camps was precise, they dispersed as soon as they detected activity. FARC rebels were ideological fighters; they were also some of the most significant cocaine traffickers in the region. In addition, much of their war in the 90s and 2000s was financed with drug funds. The US government deemed FARC a terrorist organization and a national security threat. However, ideologically, FARC did not pose any imminent threat to the US. Based on Latin America's historical perspective, the US intervention in countries run by the left-wing government was somewhat justifiable policy-wise. Drug trafficking became a problem in the late eighties for the forthcoming US administrations, and it has been a tenet of the US war on international narco-trafficking since.

To deter and decrease insurgent activity, the US provided helicopters and tested the intelligent bomb for the first time in history. The smart bomb could only be used with US approval, and the Colombian Army could not detonate it. The Pentagon held the "codes" to the intelligent bombs as a secret -for fear that the bombs could have been misused to eliminate political opponents or for any other faulty intentions.

Like any other foreign policy matter in the US, Colombia's support has had ups and downs depending on the administration. While the assistance decreased in the mid-nineties, it resumed and boosted, beginning the new millennial with Plan Colombia. The Clinton administration reduced the aid in 1996 – yet in 1998, it responded to an offensive by the rebels with a multimillion-dollar operation. Sources from Bogota claimed that former Green Berets and military fighting in the Gulf War were redeployed to combat the guerrillas. [176]

The role the US played in shifting Colombian politics cannot be diminished or denied. For decades, military assistance created more prepared military leaders with clear strategies and objectives. These strategies became paramount in decreasing FARC's capacities and operations. The US doctrine in Colombia, as the USSOF claims, has been capacity building. While that is accurate, it does not mention the direct or vicarious impact the covert operations had on the most vulnerable population. It reveals that the US intervention in Colombia shaped their domestic policy and changed the power-sharing agreement between the two longest-running political parties. [177]

The DEA's role in Colombia

The declassified documents and testimonies in the US Congress and the Department of State give a panorama of the scope of the conflict.

The Drug Enforcement Agency, especially in the early nineties, played a critical role in the conflict.

The political situation in Colombia was as complex as the conflict itself. The many fronts of the war demanded various agencies and specific strategies. The Colombian state and the paramilitary extension operated in the zones where the insurgencies had established a shadow government. The unconventional war demanded conventional and unconventional tactics.

The drug cartels had established an entire network for cocaine production, distribution, profits, and money laundering. The system was well-protected and had many layers of security. The state could not infiltrate the drug cartels and extract information from its agents. Plan Colombia covered gaps in the national security system and sectors where the state lacked strength.

The Turbay administration marked the first efforts to combat drug trafficking. Some of its policies included a Statute of Security - in August 1978. The policy included rigorous strategies against the guerrillas and the drug cartels. Furthermore, - in 1978, the Turbay administration initiated negotiations for a possible extradition treaty between Colombia and the US. The treaty - if passed - permitted the extradition of Colombian nationals to the US based on drug-related crimes.[178]

In the 1980s and early 1990s, the US Government deemed the drug problem a national epidemic. However, drug trafficking from Colombia was not the only problem. The demand for cocaine, heroin, and marijuana increased the smuggling of illegal substances from Colombia, Mexico, and the Caribbean. Despite the information campaign from the media in cooperation with the government about the damage that drugs cause to the human brain and body, various health agencies emphasized the war on drugs as a battle against an epidemic - more so than a Department of Justice problem.[179]

Challenging the Power Elite

Building modern states in Latin America is a concept that has been introduced previously and has been a challenging process. First, the state is a political and legal entity with specific characteristics in the mere definition. A modern state should have and be recognized as having a border - a body of citizens, a central government, and sovereignty – both internal and external. Second, states have a monopoly over the use of force. That is the government's response to providing security and peace within the state. Lastly, states are recognized by each other as political and legal entities. They are, of course, states that do not recognize each other, and that is problematic – especially in international relations. Yet this has not been the case in Latin America, nor have territorial claims been prevalent. The problem in many Latin

American countries stems from poor governance, political violence, intolerance, dictatorships, military coups, revolutions, and ideological warfare.

Post-colonial states have always presented problems, and in many cases, the approach to view the failed state (or a post-colonial state) must look at a few indicators. First, the state's idea must be similar to all contemporary states. Secondly – the concept of the state must have a certain degree of importance, whether a failed state or not, and thirdly, the relationship between the state and its practices must be the focus of the approach. Thus, instead of merely comparing actual states to the underlying assumptions (on which they function), we *should focus on how states are shaped by the practices of various actors and by their interrelationships and interactions.* [180]

Hence, the idea of the state is overpowered by the practices, behaviors, and actions of the state actors rather than theoretical frameworks. Comparative politics cover states across the board, and to make assumptions about the nature of certain states, one must compare them to others. In cases where a state is on the verge of collapse, it is necessary to look at practices that have brought the state failure and match them to that of a normal functioning state. Thus, the comparative analysis between states, and more importantly, between failing states vs. normal states, will help the argument in the case of Colombia. [181]

Institutional Crisis and the Iron Fist

The Colombian conflict precedes the formal founding of the FARC in 1964. Colombian political and social life has always suffered from polarization and violence based on significant wealth inequalities suffered by landless peasants in the periphery. [182] After a brief military rule (1953-1958), the two major parties, the Liberals and Conservatives – arranged a power-sharing agreement and formed a National Front coalition. Under the agreement, the parties will rotate the power every four years. This power-sharing agreement lasted for sixteen years, yet it did not settle the disagreement between the extremist factions within the parties. [183] The botched efforts to create stability – drove the left-wing factions to launch two insurgencies: The FARC and the ELN.

The elite theory helps us understand that the gap between the ruling class and the masses does not threaten the elites as long as all strata and classes widely accept the status quo. Yet, suppose any of the lower classes question the status quo, the social order, and the political developments. In that case, the elites will use the state's power to crush any rebellion or organizations, legal or illegal, that would threaten its existence. In that case, the ruling class will not agree to any power-sharing with political forces that rise from the bottom without a revolution. In a failed state, the elite fights for its existence, legitimacy, and economic, military, and judicial control. The elite will put all the necessary means into its disposition, significantly the military might, to restore order. Such were the efforts to create a militia force- separate yet in conjunction with the military to carry out an operation that the army opted out of.

In the chapter about the narco-traffickers, I touched upon the institutional failure when M19 guerrillas took over the Justice Palace for the mere reason that the attack -allegedly- planned by the Medellin Cartel sought to destroy evidence related to various crimes. However, this chapter will explain how the political elite used its military strength to protect the institutions and the political class. November 6, 1985, marks a turning point for the Colombian political elite.

Summary of the Siege

On November 6 and 7, 1985, commandos from M19 violently entered Colombia's Courthouse in Bogota, took over the building, opened fire on police and security personnel, and captured as hostages - judges, magistrates, employees, and other people who were in the courthouse. The M19 guerrillas, led at the time by Commander Luis Otero, got to the building's fourth floor and other strategic points. Among the hostages was the President of the Supreme Court, Chief Justice Alfonso Reyes Echandía. [184]

The assailants took over the building. At least a dozen people who vanished were heard of, or their bodies were never recovered. They were perhaps burnt or destroyed in the fire. The operation was swift, and surprisingly – as witnesses say – there had been prior warnings that the guerrillas would take over the Palace justice. Still, little is known if the state took the threats seriously.

Commandos took about 300 people hostages, including the 24 justices and 20 other judges, 12 of whom died. They first seized the Supreme Court Justice and the Presidency of the Constitutional Court called Sala Constitucional; Manuel Gaona Cruz, a key member, was assigned to deliver an opinion on the constitutionality of the extradition treaty between the United States and Colombia. Both Alfonso Reyes Echandia and Manuel Gaona Cruz were assassinated.

The State's Response was one of the fiercest interventions. The Colombian military killed many civilians and guerrillas in the attack. There were speculations that the magistrates were killed by the Colombian army, which the Colombian State denied, but it compensated the families of the deceased magistrates. About three hours after the initial seizure, army troops rescued about 200 hostages from the lower three floors of the building; the surviving gunmen and remaining hostages occupied the upper two floors. A recording was delivered to a radio station soon after the seizure, saying that the M-19 group had taken over the building "in the name of peace and social justice."[185] From the Supreme Court, the M-19 members demanded President Belisario Betancur come to the Palace of Justice to stand trial and negotiate. The president refused and ordered an emergency cabinet session. [186]

> *"The Palace of Justice was a naked story, a theatrical performance in public space in which the actors naturally and spontaneously overreacted as if no one were watching (...). Everything was done to "defend democracy, teacher," said one; "There are still men willing to die for a cause," chanted the other. And the country, astonished, listened to them (Vélez and Atehortúa, 1993: 291).*[187]

The thirty-five M19 guerrillas[188] who stormed from the basement of the building with guns and ammunition and seized the Supreme Courthouse marked the ultimate institutional crisis in Colombia's modern history or any modern state. The armed commando detonated explosives for two days, shot indiscriminately, and took hostage justices, court clerks, employees, auxiliaries, secretaries, and civilians.[189] The assailants demanded that the magistrates try President Belisario Betancourt, who refused to negotiate with the guerrillas. The demands were unrealistic legally, but politically, they destroyed the reputation of the Colombian state, the institution of justice, and the presidency. Caught in a national crisis, Betancourt released a military plan to retake the Justice Palace with force. On November 7, the military led by General Jesús Armando Arias Cabrales, commander of the Thirteenth Army Brigade in Bogotá, surrounded the building with EE-9 Cascavel armored cars and knocked down the entrance of the Palace. The attack's aftermath revealed a disaster. Twelve magistrates were killed and, among others, staff and civilians. The armored tanks launched rockets, bazookas, and other unidentified weapons – declared Colombian National TV. The Inter-American Commission on Human Rights, which called it a massacre led by the state, criticized the handling of the situation. Belisario Betancourt assumed responsibility for the inadequate handling of the crisis and expressed his sorrow for the victims. Yet, he claimed his duty as President was to protect the institutions. Thus, he justified the violence in the name of the state.

Political violence:
Assassination of presidential candidates - 1989

Between 1989- and 1990, while campaigning for the presidential race, three prominent figures were assassinated while campaigning: Carlos Pizarro Leon

Gomez, Luis Carlos Galan Sarmiento, and Bernardo Jaramillo Ossa.

The assassinations took place while the candidates were under the protection of DAS,[190] who is alleged to have had ties with paramilitary groups. "Colombian Commission of Jurists and the Lawyers Collective "José Alvear Restrepo" had filed a petition to the Interamerican Commission of Human Rights alleging that the crime against Carlos Pizarro[191] was organized by the paramilitary groups – in cooperation with the DAS.

Bernardo Jaramillo Ossa – also a presidential candidate, was killed while traveling to Santa Marta with his wife.

In 1989, political violence culminated with the assassination of Carlos Luis Galan Sarmiento, a prominent Colombian politician who ran for President. Galan was a career politician and founded a movement against mainstream politics represented by the traditional parties; he faced opposition from politics and organized crime. Galan was quite eloquent and charismatic. He was part of the political elite and clearly understood Colombia's future, yet his vision was partial.

Galan believed in the rule of law and waged war against the narco-traffickers, blaming them for the ramping violence. He was a proponent of the extradition law, and if elected, it posed a threat to the drug cartels and their illegal activity. Yet, the politics in Colombia were more complex than they appeared. With the rise of the neo-liberal movement, the balance of power would shift more to the left. Even though neo-liberalism was a modern faction of the Liberal Party, it posed a threat and weakened its position in domestic politics.

The popularity of Galan rose in 1989 with his rhetoric against the Medellin Cartel, which became the centerpiece of his legislative agenda. He promulgated a reformed judicial system – as his crime-fighting strategy. Galan seemed to overlook the deep divide that drove the country to a prolonged conflict and focused only on organized crime. His popularity was on the rise, especially in urban areas. He had practically no influence or strategy for rural Colombia, where the conflict was concentrated. Galan's foes came from the darkest corners of Colombian politics, elites and gangs. His politically motivated assassination shows the deep crisis, the intolerance, and the weak state of affairs in Colombian politics. It was a turning point for the political parties to reassess their position. The right-wing parties had the upper hand in the conflict and the support of the paramilitary along with a costume-made national security doctrine that worked only for the right wing. In conclusion, the three presidential candidates assassinated in 1989 had left-leaning political views, and allegedly, the paramilitary was involved either in planning or executing the attacks.

The Strategy to Prevent State Collapse

To win a conflict, one must first wage war and assess the enemy's military capacity, zones of operation, and capabilities to overthrow a government. The war must have a just cause embodied in a doctrine. The Colombian government launched a witch-hunt campaign against the left and its sympathizers in higher institutions in the late nineties and early seventies.

The National Doctrine: Alvaro Uribe (2002) undertook several measures to strengthen national security after being elected president. He implemented a policy called "Democratic Security." The policy mobilized resources for war to take control of institutions, infrastructure, and territories under FARC control. The support of the U.S. military and counterterrorism played an important role here. Uribe was able to mobilize the Colombian Army and regain state control over critical areas of the country.[192]

Outsourcing military powers to non-statal actors: Even though the paramilitary were demobilized and shunned by human rights activists for their crimes, they bore one of the most significant responsibilities during the conflict. Along the conflict, the paramilitary morphed from a non-state agent providing security to private companies to a state-sponsored terrorist organization. Later, with the

Castaño brother's outrage, the paramilitary sought to enter the political scene.

> *Right-wing paramilitary forces, often in coordination with conservative political and business leaders, organized to fight FARC units and their supporters directly. They also targeted social leaders, union representatives, journalists, politicians, and other political actors who challenged the conservative order.*[193]

Uribe successfully ran for reelection in 2006. During his second term, Uribe struck a deal with the paramilitary. During 2003 and 2006, more than 30,000 fighters demobilized.

> *These groups, operating under the banner of the United Self-Defense Forces of Colombia (AUC), reached a deal in 2003 to begin to disarm. In 2005, they accepted the government's controversial offer of reduced jail terms and protection from extradition.*[194]

Many were pardoned or received a reduced sentence for the grave war crimes. Hence, impunity reigned supreme, and the influence of the right-wing militia in politics is quite strong even nowadays.

A clear view of the political and social situation in Colombia is revealed in the Human Rights Report. The findings reveal a panorama of Colombia that perhaps is less problematic than it was during the armed conflict but reflects what the Colombian government is still incapable of doing and how the political game has placed the state in a fragile position. Furthermore, the government continues its abuses, human rights violations, arbitrary killing, and extrajudicial executions. State agents reported thousands of false positive cases- of arbitrarily executed civilians under the excuse that they were FARC warriors.

The gross human rights violations and killings by the Colombian state continue despite the ongoing investigations for war crimes. The impunity and the slow investigations on accountability and shared responsibility have characterized the process, and moving forward seems farfetched. Between 2008 and 2020, army soldiers and, generals, police officers and have been involved in numerous reported crime investigations. The JEP (Special Jurisdiction for Peace) warns that an escalation of the conflict in search for ELN or another guerrilla grouping can also increase the civilian casualties. President Ivan Duque has denied any state involvement and extrajudicial killing during his years of presidency, stating that their policy and doctrine surrounding the facts were in alignment with human rights. Yet reports from the UN and other agencies tell a different account (*see footnote*). [195] A*ccording to the Human Rights Report in 2020:*

> *Significant human rights issues included Reports of unlawful or arbitrary killings, reports of torture and arbitrary detention by government security forces and illegal armed groups… killings, and other violence against trade unionists.*[196]

To summarize the argument, Colombia is still fragile even after the peace agreement. Opposition to the deal has shown a lack of willingness to abide by the terms, especially the integration of demobilized FARC fighters into social and political life. Lack of opportunities has driven many former fighters to the outskirts of Colombia, seeking to reorganize in the areas where the government presence is not as strong.

Colombia continues to be a case of toxic politics and is now in a very fragile political position. The peace process is an imaginary state of politics ready to turn sour any day. Ostracizing former guerrillas and strengthening state-sponsored violence will prompt social unrest, more violence, and potentially an armed conflict. By definition, Colombia can be considered a failed state- since it does not control its entire national territory along with other indicators of the fragile state index. Even though smaller in scope and operations, the drug cartels continue to export tons of cocaine in drug markets in Europe and the US.

At face value, Colombia seems to be a democratic society, yet there is much more to democracy than seemingly free elections. Democratic institutions will cease if the state does not function as a security instrument for its citizens and constituents. Likewise, if the state uses its monopoly over violence to fulfill the interests of one political group or entity, it is destined to fail.

This study analyzes the state of Colombia on the verge of collapse with the utmost objectivity. It highlights the roots of the conflict, the outcome, and the key players. Colombia gained international attention due to the guerrilla war, drug cartels, and mostly the kidnapping of civilians by the guerrillas. Geographically – Colombia is rich in arable land, vegetation, and mountainous terrain. The country is also known as the Republic of Colombia, which is situated on the South American continent and borders Brazil, Venezuela, Ecuador, Peru, and Panama. It has approximately 55 million people, and the official currency is el Peso Colombiano. The country's capital is Bogota, where the political elite and the government operate.

Colombia has vast arable lands and products and has a diverse population varying from mixed-race (Los mestizos) to Euro-Colombianos, Afro-Colombianos, and indigenous

Colombians. Throughout the book, I covered how the conflict changed society's dynamics and fabric. It polarized the population in two extremes according to their political affiliation. Geography also played a role in determining loyalties to the right-wing militia and corporate or the left-wing guerrillas. As the world watched the conflict unravel, the fighting parties committed some of the most outrageous human rights violations.

The complexity of the conflict is mainly attributed to key actors who entered the political game and benefitted from it, or the game turned to their demise. Yet, the public learned that the political elite survived despite what theoretical templates explain. For example, Colombia hardly had experienced any military coup, except Rojas Pinilla's regime, which claimed to restore order amongst the fighting parties when the violence erupted and got out of control. As discussed earlier, military coups are less costly than revolutions and do not displace larger polities. Moreover, the military juntas can be overthrown and replaced by the traditional elite.

> *Colombia's elites have been able to repeatedly block attempts to redistribute land or carry out significant political reform. The country largely avoided the military coups and economic crises that beset its neighbors in the 20th century. Still, this stability helped store up a legacy of unresolved*

social problems. Colombia is "the only Latin American country in which the traditional parties and elites neutralized all political reform efforts." [197]

The conflict had roots in the lack of governing fundamentals and democratic values, the failure to accept people's will, and the lack of rigor to serve the people. Hence, the priority of the elite, despite their political affiliation, was to maintain power. The war highlighted many social wounds and phenomena. It showed how rotten a post-colonial society was and how deep colonialism was rooted in people's mentality that the elite failed to build a modern and robust state even after independence. As a result, the political aristocracy and the economic oligarchy created their network of paramilitary and militia to protect only their interests. The strategy proved disastrous for the people, but it shielded the wealthy landowners from various existential threats - mainly from guerrillas and other left-wing insurgencies.

Implications for the Elite Theory

In the book's introduction, I presented a short overview of the critical concepts of elite theory. In congruence with the literature, we see that the Colombian elite derived from a class of wealthy landowners who pushed for self-governing and did away with the idea of Grand Colombia.

Furthermore, this class emerged from one of the most rigid colonial systems without prior experience in nation and state building and inherited very few technocrats to maintain some normalcy in the country's governance, keep the finances, and build a state treasury. The governing elite kept many of the colonial policies it inherited from the metropolis regarding the judicial branch and legislating.

It would be challenging to chronologically describe the evolution of the elite and its governing power – since the post-colonial states do not follow the traditional governing pattern and state-building frameworks of the Western democracies. Hence, despite the corruption and instability, what we observe in the newly formed states now is tremendous pressure from international actors and institutions to create a balance of power by allowing these states some assistance in nation-building. Latin American countries seldom received any international aid upon creation. Many systems evolved as products of dictatorships, regimes, military juntas, and revolutions. In Colombia, the events took the wrong turn during the years of violence, which was followed by an armed conflict. First, the division among the political parties pulled people to the extremes and caused a bloody war for at least five decades. Unlike democratic societies where the elite is characterized by a peaceful transfer of power, in authoritarian regimes, the elite falls into other normative, i.e., a limited or a non-existent provision of democratic processes. Yet at face value – and until recently - the Colombian elite provided some indicators of democratic processes such as free elections.

It was not until the power-sharing agreement in 1958-1974 between the conservatives and the liberals to rotate the power to oust Gustavo Rojas Pinilla's regime. The pact was named the National Front Pact – and again demonstrated that the Colombian elite was one solidified class with strong powers and a will to rule the masses despite ideological differences.

The elite theory sees the political system divided into two groups: First, there is the elite or the political entrepreneurs with their demagoguery and ideologies, simultaneously manipulative and skilled in the political game. On the other hand, many people depend on the elite to make political decisions to draft and pass laws and policies. The people might show interest in these processes but be passive and not get involved in the country's politics. The indifference sometimes goes as far as not participating in the voting process.[198] In authoritarian or semi-authoritarian regimes where elections are often contested – this might not be a problem, but for consolidated democracies, contesting or mass boycotting - becomes problematic.

Thus, depending on the political system, the people rely primarily on the political elite to ensure that the system functions in all government branches – at least at the central and local levels.

Hence, governing involves the risk of the system failing due to outside or domestic threats, and the elite faces opposition and struggles to maintain power.

In a democratic system, the elite has to invent a narrative that fits the masses and their concerns. Otherwise, voters will drift away from specific candidates and elect the opponent or boycott the voting. It is worth mentioning that in democracies, people are indirectly involved in the decision-making process, which is done through formal or informal institutions. Ultimately, the elite can make decisions regardless of what the people might approve or disapprove of. [199]

According to the elite's theory, increased participation in the democratic system will threaten its stable yet functioning nature. Thus, the fewer people participate in the country's political life, the longer the status quo remains. That is, elites fear massive and ardent direct participation because it will shake the balance of power and power-sharing agreements and destabilize the traditional power transfer.

Hence, as I mentioned above, the Colombian elite has a different nature and many characteristics that distinguish it from the traditional elites. For that matter, it hardly fits in the theory of the elites. We can conclude that politically, it behaved like banditry until after the years of violence (the late fifties and early sixties) when the country sought stability. Analyzing the Colombian elite using the above format, we can conclude that it sought to advance its agenda. Contrary to the democratic processes and values, - the Colombian elite did not serve; it only benefitted the system it created. What the theory postulates is entirely predictable. The driving of the people to the extreme ideologies brought almost to the collapse of the system, the contested elections, the assassination of political leaders and candidates, and the fifty-year-long conflict.

What is good for the elite should suit everyone; this holds some veracity. In this study, I was able to give an objective view of the conflict. I was able to point out differences between how Colombian traditional elites evolved throughout the decades or even centuries and how elite outcasts end up outside the political discourse, forming insurgencies, rebel groups, or paramilitary structures to preserve the status quo or come to power.

I showed the parties involved in the conflict, pointing out that a few factors, such as the Power Elite, state and institution, government armed forces, and non-government armed forces, will serve the ruling class and preserve the status quo despite ideological differences as long as they remain in the political debate.

Lastly, militarizing the state is one method elites use to preserve power. Latin America has been notorious for military coups, juntas, and dictatorships from the 1960s until the early 1980s. One factor that led to these nontraditional transitions of power was militarizing the state by increasing military personnel and distributing funds to boost the army, police, and other armed forces. In the case of Colombia, strengthening the military and police force was the only way to prevent a state collapse.

In conclusion, the protracted conflict in Colombia was a survival struggle for the political elite and wealthy landowners. The elite usually advances the country's economic and political agenda but builds a solid nation-state to confront internal and external threats. The Colombian elite was not able to build a strong state. It was a failed state on the verge of collapse during the conflict.

The elite was divided into various factions that moved to the extremes. The elite mentality remained local or regional since Grand Colombia. It was never unified for the greater good. The liberals – and the Communist Party were also part of the political elite, and FARC was the extremist faction that challenged the state militarily – on the other hand – the conservatives, represented by the wealthy landowners, used the paramilitary to combat the left-wing insurgencies. At the same time, the traditional elite militarized the state to establish order and prevent a collapse.

Bibliography

2014, Autor de Kaos. "Colombia: A 20 Años de La Muerte Del Camarada Jacobo Arenas, Su Biografía." Kaos en la red, August 11, 2011. https://archivo.kaosenlared.net/colombia-a-20-a-os-de-la-muerte-del-camarada-jacobo-arenas-su-biograf-a/.

Archila, Mauricio. "Violencia y Política En Colombia. De La Nación Fragmentada a La Contrucción Del Estado | Anuario Colombiano de Historia Social y de La Cultura." repositorio.unal.edu, 2003. https://revistas.unal.edu.co/index.php/achsc/article/view/17115.

Arjona, Ana. *Rebelocracy: social order in the Colombian civil war*. Cambridge, U.K: Cambridge University Press, 2017.

BANDITRY AND INSURGENCY IN COLOMBIA - The World Factbook. Accessed September 20, 2023. https://www.cia.gov/readingroom/docs/DOC_0000598514.pdf.

"Banditry and Insurgency in Colombia." CIA.GOV, June 2001.

Beaubien, Jason. "Bombs in Baby Food Jars Are Just One Part of Colombia's Land Mine Problem." JMU, January 10, 2018. https://www.jmu.edu/news/cisr/2018/01/10-colombia.shtml.

Biden, Joe. " AID TO ``PLAN COLOMBIA'': THE TIME FOR U.S. ASSISTANCE IS NOW." https://www.govinfo.gov, 2000.

Bonilla, Daniel. "Liberalism and Property in Colombia: Property as a Right and Property as a Social Function."

Fordham Law Review 80, no. 3 (December 2011). https://doi.org/10.1007/978-981-13-7189-9_8.

Brett, Roddy. "Elite Bargains and Political Deals Project: Colombia Case Stud." Stabilization Unit, February 2018. https://assets.publishing.service.gov.uk/government/uploads/system/uploads/attachment_data/file/766016/Colombia_case_study.pdf.

Briscoe, Charles H. "Across the Pacific to War." Across the Pacific to War: The Colombian Navy in Korea, 1951–1955, 2006. https://arsof-history.org/articles/v2n4_across_pacific_page_1.html.

Byman, Daniel L., Peter Chalk, Bruce Hoffman, William Roseneau, and David Brannan. Trends for outside support for insurgent movements. Accessed September 12, 2023. https://www.rand.org/dam/rand/www/external/congress/terrorism/phase2/insurgent.pdf.

Cadavid, Erich Saumech. Historia de la guerrilla en Colombia. Accessed September 20, 2023. https://www.didacticamultimedia.com/registro/estudios/10/documentos/guerrilla_colombiana.pdf.

Castaño, Carlos, and Mauricio Aranguren Molina. *Mi Confesión: Carlos Castaño Revela Sus Secretos*. Bogotá: Oveja Negra, 2002.

Chomsky, Noam. *Rogue states: The rule of force in World Affairs*. London: Pluto Press, 2016.

Collier, Paul, and Anke Hoeffler. Greed and Grievance in Civil War. Accessed September 24, 2023. https://documents1.worldbank.org/curated/en/192611468769173749/pdf/multi-page.pdf.

"Colombia CJA Litigation Cabrera v. Jiménez Naranjo Urán, et al. v. Plazas Vega." CJA. Accessed April 24, 2024. https://cja.org/where-we-work/colombia/.

Colombia: Background and U.S. Relations - CRS Reports. Accessed September 19, 2023. https://crsreports.congress.gov/product/pdf/R/R43813.

Colombia reports. "Cali Cartel Gave $10m to Samper's Presidential Campaign: Cartel Executive." Colombia News | Colombia Reports, February 26, 2015. https://colombiareports.com/amp/cali-cartel-gave-10m-to-sampers-presidential-campaign-cartel-executive/.

Colombia reports. "FARC Claim Responsibility for Killing One of Colombia's Most Legendary Politicians." Colombia News | Colombia Reports, October 5, 2020. https://colombiareports.com/amp/farc-claim-responsibility-for-killing-one-of-colombias-most-legendary-politicians/.

Colombia reports. "The Wars for Colombia's Cocaine Containers: Part 3: Santa Marta." Colombia News | Colombia Reports, April 25, 2022. https://colombiareports.com/the-wars-for-colombias-cocaine-containers-part-2-santa-marta/.

Covert action in Colombia | The Washington Post. Accessed April 16, 2024. https://www.washingtonpost.com/sf/investigative/2013/12/21/covert-action-in-colombia/.

Cragin, Kim, and Bruce Hoffman. Arms trafficking and Colombia - rand corporation. Accessed September 24, 2023. https://www.rand.org/content/dam/rand/pubs/monograph_reports/2005/MR1468.pdf.

Cramer, Christopher, and Elisabeth Jean Wood. "Introduction: Land Rights, Restitution, Politics, and War in Colombia." *Journal of Agrarian Change* 17, no. 4 (September 20, 2017): 733–38. https://doi.org/10.1111/joac.12239.

"Declassified Documents Constitute Key Evidence in Judgement against Colombian Paramilitary." National

Security Archive, December 24, 1997. https://nsarchive.gwu.edu/briefing-book/colombia/2021-10-04/declassified-documents-key-judgment-against-colombian#:~:text=Washington%2C%20D.C.%2C%20October%204%2C,judgment%20against%20a%20former%20Colombian.

Defense Intelligence Agency > FOIA > Foia Electronic Reading Room > Foia Reading Room: Colombia. Accessed April 27, 2024. https://www.dia.mil/FOIA/FOIA-Electronic-Reading-Room/FOIA-Reading-Room-Colombia/FileId/39781/.

Defense Intelligence Agency > FOIA > Foia Electronic Reading Room > Foia Reading Room: Colombia. Accessed April 27, 2024. https://www.dia.mil/FOIA/FOIA-Electronic-Reading-Room/FOIA-Reading-Room-Colombia/FileId/39781/.

DeMare, Brian James. *Land wars: The story of China's Agrarian Revolution*. Stanford, CA: Stanford University Press, 2019.

"Department of Justice: Homepage: United States Department of Justice." Department of Justice | Homepage | United States Department of Justice, September 9, 2016. https://www.justice.gov/.

DeYoung, Karen. "Politics - The Washington Post." Washington Post, August 31, 2000. https://www.washingtonpost.com/politics/.

"The Drug Enforcement Administration (DEA) 1980-1985." The Drug Enforcement Administration. Accessed April 26, 2024. https://www.dea.gov/sites/default/files/2021-04/1980-1985_p_49-58.pdf.

The Drug Enforcement Administration (DEA) 1990-1994. Accessed April 24, 2024. https://www.dea.gov/sites/default/files/2021-04/1990-1994_p_67-76.pdf.

DRUG ENFORCEMENT ADMINISTRATION 1990-1994, 1994.

Dunleavy, Patrick, and Brendan O'Leary. *Theories of the state: The Politics of Liberal Democracy*. Basingstoke: Macmillan Education, 1987.

Fragile State Index Methodology and CAST Framework, 2017. https://fragilestatesindex.org/wp-content/uploads/2017/05/FSI-Methodology.pdf.

Francisco, Alvaro, Amaya Villarreal, and Gustavo Emilio Cote-Barco. Vista de la Toma del Palacio de Justicia: La reparación del daño en eventos de violación de derechos humanos, Winter 2006. https://revistas.javeriana.edu.co/index.php/vnijuri/article/view/14652/11821.

"The Friends of 'El Viejo': Declassified Records Detail Suspected Paramilitary, Narco Ties of Former Colombian President Uribe." National Security Archive, September 23, 1991. https://nsarchive.gwu.edu/briefing-book/colombia/2020-08-31/friends-el-viejo-declassified-records-detail-suspected-paramilitary-narco-ties-former-colombian.

Fulo Regilme Jr, Salvador Santino. "American Foreign Aid and Colombia's Human Rights Tragedy." E International Relations, July 7, 2018.

Guide to the analysis of insurgency, January 5, 2009. https://irp.fas.org/cia/product/insurgency.pdf.

Gutierrez Sanín, Francisco, and Mauricio Baron . Re-stating the state in Colombia - LSE, 2005. https://www.lse.ac.uk/international-development/Assets/Documents/PDFs/csrc-working-papers-phase-one/wp66-restating-the-state-in-colombia.pdf.

Gutierrez Sanín, Francisco. "Tough Tradeoffs: Coca Crops and Agrarian Alternatives in Colombia." International Journal of Drug Policy, February 18, 2021. https://www.sciencedirect.com/science/article/abs/pii/S0955395921000542.

Hartlyn, Jonathan. "Drug Trafficking and Democracy in Colombia in the 1980s ..." https://www.icps.cat, May 1989. https://core.ac.uk/download/pdf/13283356.pdf.

Harvey, Tom. "Colombia Close to Capturing 'Public Enemy No. 1' - UPI Archives." UPI, May 20, 1990. https://www.upi.com/Archives/1990/05/20/Colombia-close-to-capturing-Public-Enemy-No-1/1859643176000/.

HEARINGS BEFORE T.rIE PERMANENT SUBCOMMITTEE ON INVESTIGATIONS OF THE COMMrrTEE ON GOVERNMENTAL AFFAIRS UNITED STATES SENATE ONE HUNDRED FIRST CONGRESS, 1989. https://www.ojp.gov/pdffiles1/Digitization/146771NCJRS.pdf.

Higgins, Hillary. "Counternarcotics to Counterinsurgency: Assessing the Transformation of US Economic Assistance to Colombia, 1998–2002." Undergraduate Thesis Conference, March 2015. https://conferences.wcfia.harvard.edu/undergraduate_thesis/publications/counternarcotics-counterinsurgency-assessing-transformation-us.

Hofmann, Claudia, and Urlich Schneckener . "Engaging Non-State Armed Actors in State and Peace-Building: Options and Strategies." *International Review of the Red Cross* 93, no. 883 (September 2011).

Hoyos - Zuñiga, Edgar Jesus. "Lucha Por La Tierra En El Resguardo Indigena Nasa de Las Delicias 1977-1986." repositoriounicauca.edu, 2019. https://pdba.georgetown.edu/Comp/Agrario/propiedad.html.

Hughes, Michelle, and Michael Miklaucic. " Colombia and the FARC: From Military Victory to Ambivalent Political Reintegration?" Essay. In *Impunity: Countering Illicit Power in War and Transition.* Washington, DC: Center for Complex Operations, 2016.

"Human Rights Watch: Informe Anual 1998 (Eventos de 1997) — Colombia." human rights watch | informe anual 1998 (eventos de 1997) — colombia. Accessed September 24, 2023. https://www.hrw.org/legacy/spanish/inf_anual/1998/colombia.html.

InSight Crime. "FARC." InSight Crime, October 20, 2022. https://insightcrime.org/colombia-organized-crime-news/farc-profile/#:~:text=The%20FARC%20are%20the%20oldest,drug%20trade%20on%20various%20levels.

Ioanes, Ellen. "What's behind Africa's Recent Coups." Vox, September 10, 2023. https://www.vox.com/world-politics/2023/9/10/23866908/africa-coup-gabon-niger-mali-burkina-faso.

Jenkins, Nash. "Colombia: More than 100 Butchered Bodies Found Beneath Jail." Time: February 19, 2016. https://time.com/4230222/la-modelo-sewer-bodies-colombia/.

"The Kingpin Strategy: Did It Work and Is It Still Relevant?" The Kingpin Strategy: Did It Work and Is It Still Relevant? | DEA Museum. Accessed April 29, 2024. https://museum.dea.gov/video-archive/kingpin-strategy-did-it-work.

"Las Farc-EP En El Caribe Colombiano." *El otro derecho*, November 10, 2020, 27–33. https://doi.org/10.2307/j.ctv1cfthhw.5.

Leech, Gary. *The FARC: The most prolonged insurgency.* London: Zed Books Ltd, 2011.

Lopez-Uribe, Maria del Pilar, and Antonio Sanchez Torres. On the agricultural origins of civil conflict in Colombia. Accessed April 12, 2024. https://www.lse.ac.uk/lacc/publications/PDFs/Sanchez-LopezUribe-OriginOfConflict-03-12-2017.pdf.

M-19 guerrilla attack on Colombian Palace of ... Accessed April 24, 2024. https://www.cia.gov/readingroom/docs/CIA-RDP91B00874R000100040006-2.pdf.

Mariotti, Claudia. "Elite Theory." The Palgrave Encyclopedia of Interest Groups, Lobbying and Public Affairs, December 9, 2020. https://www.academia.edu/44666132/Elite_Theory.

Marx, Karl. "Chapter 26 The Secret of Primitive Accumulation." *Capital*, April 17, 2008. https://doi.org/10.1093/owc/9780199535705.003.0034.

Maullin, Richard L. Defense Technical Information. Center Soldiers, Guerrillas and Politics in Colombia. Accessed 20AD. https://apps.dtic.mil/sti/pdfs/ADA389095.pdf.

Medina, Oscar. "Cocaine Is Expected to Overtake Oil as Colombia's Main Export This Year." Bloomberg.com, September 14, 2023. https://www.bloomberg.com/news/articles/2023-09-14/cocaine-is-set-to-overtake-oil-to-become-colombia-s-main-export.

Miroff, Nick. 'Plan Colombia': How Washington learned to love Latin American intervention again - The Washington Post, September 6, 2006. https://www.washingtonpost.com/world/the_americas/plan-colombia-how-washington-learned-to-love-latin-american-intervention-again/2016/09/18/ddaeae1c-3199-4ea3-8d0f-69ee1cbda589_story.html.

"MMP: Revolutionary Armed Forces of Colombia (FARC)." CISAC. Accessed September 21, 2023.

https://cisac.fsi.stanford.edu/mappingmilitants/profiles/revolutionary-armed-forces-colombia-farc.

Moyar, Mark, Hector Pagan, and Wil R Griego. Persistent E N Gagem Ent in Colom BIA - Washington Office on ..., July 2014. https://www.wola.org/files/141113_jsou_notes.pdf.

Multinational Operations - Joint Chiefs of Staff. Accessed September 19, 2023. https://www.jcs.mil/Portals/36/Documents/Doctrine/pubs/jp3_16.pdf.

Nctc. "The Methods and the Tactics of Global Terrorism." Counterterrorism Guide - National Counterterrorism Center. Accessed January 19, 2024. https://www.dni.gov/nctc/index.html.

Ospina, Hernando Calvo. *Colombia, Laboratorio de embrujos democracia Y terrorismo de estado*. Tres Cantos, Madrid: Foca, 2008.

"An Overview of Conflict in Colombia." International Center for Transitional Justice, 2009. https://www.ictj.org/sites/default/files/ICTJ-Colombia-Conflict-Facts-2009-English.pdf.

"The Palace of Justice Attack – Losses and Gains." https://nsarchive2.gwu.edu, December 9, 2008.

Palacios, Marco. *Between legitimacy and violence: A history of Colombia, 1875-2002*. Durham: Duke University Press, 2007.

Pappier, Juan. "Left Undefended." Human Rights Watch, August 17, 2023. https://www.hrw.org/report/2021/02/10/left-undefended/killings-rights-defenders-colombias-remote-communities.

Paramilitarismo Balance de la contribución del CNMH al esclarecimiento histórico, 2018. https://www.centrodememoriahistorica.gov.co/micrositios/balances-jep/descargas/balance-paramilitarismo.pdf.

"PENSAMIENTO MARXISTA COLOMBIANO Y ORIGEN Y PERMANENCIA DE LAS FARC-EP." usergioarboleda.edu, January 2016.

Piccone, Ted. The Colombian experience with Transitional Justice, July 2019. https://www.brookings.edu/wp-content/uploads/2019/06/FP_20190708_colombia.pdf.

Press, The Associated. "Judge Investigating Escobar Is Shot to Death in Colombia (Published 1992)." The New York Times, September 19, 1992. https://www.nytimes.com/1992/09/19/world/judge-investigating-escobar-is-shot-to-death-in-colombia.html.

Priest, Dana. "Covert Action in Colombia." The Washington Post. Accessed September 23, 2023. https://www.washingtonpost.com/sf/investigative/2013/12/21/covert-action-in-colombia/.

Rabasa, Angel, and Peter Chalk. *Colombian labyrinth: The synergy of drugs and insurgency and its implications for regional stability*. Santa Monica, CA: RAND, 2001.

Restrepo, José Alvear. "The Contribution of Chiquita Corporate Officials to Crimes against Humanity in Colombia." International Human Rights Clinic, 2017. https://www.fidh.org/IMG/pdf/rapport_chiquita.pdf.

Richani, Nazih. *Systems of violence: The Political Economy of War and peace in Colombia*. Albany: Suny Press, 2013.

Richburgh, Keith B. "Reagan Order Defines Drug Trade as Security Threat, Widens Military Role - The Washington Post." Washington Post, 1986. https://www.washingtonpost.com/archive/politics/1986/06/08/reagan-order-defines-drug-trade-as-security-threat-widens-military-role/309fdc6f-e5b8-4a64-8249-7b51182b3db1/.

Rinehart, Christine Sixta. *Volatile social movements and the origins of terrorism: The Radicalization of Change.* Lanham, MD: Lexington Books, 2013.

Roman, Juan F. "Guerrilla Violence in Colombia: Examining Causes and Consequences." Calhoun Home, June 1, 1994. https://calhoun.nps.edu/handle/10945/42913.

SALAZAR, ALONSO. *La Parabola de Pablo*. DEBOLSILLO, 2019.

Seagrave, Alan. "Conflict in Colombia: How Can Rebel Forces, Paramilitary Groups, Drug Traffickers, and Government Forces Be Held Liable for Human Rights Violations in a Country Where Impunity Reigns Supreme, Alan Seagrave (2000)." International Centre on Human Rights and Drugs Policy. Accessed April 12, 2024. https://www.hr-dp.org/contents/646.

Self, Kevin A. COUNTERTERRORISM POLICY IN COLOMBIA. Accessed September 23, 2023. https://apps.dtic.mil/sti/pdfs/ADA389095.pdf.

Staff, Reuters. "Colombia Army Raids Illegal Mines Funding FARC Rebels." Reuters, May 11, 2015. https://www.reuters.com/article/colombia-mining-illegal-idINKBN0NW27820150511.

Stone, Hannah. "Colombia Elites and Organized Crime: Introduction." InSight Crime, October 22, 2023. https://insightcrime.org/investigations/colombia-elites-and-organized-crime-introduction/.

"A Supreme Court Judge Whose Predecessor Was Assassinated Was... - Upi Archives." UPI, October 30, 1986. https://www.upi.com/Archives/1986/10/30/A-Supreme-Court-judge-whose-predecessor-was-assassinated-was/9633531032400/.

"Syria: The Story of the Conflict." BBC News, March 11, 2016. https://www.bbc.com/news/world-middle-east-26116868.

"Third International." Encyclopædia Britannica. Accessed September 12, 2023. https://www.britannica.com/topic/Third-International.

Thu, Lee TaylorLee Taylor is a Georgetown University School of Foreign Service student and AllenWalter. "Case Analysis: The FARC in Colombia." Case Analysis: The FARC in Colombia | Small Wars Journal. Accessed September 12, 2023. https://smallwarsjournal.com/jrnl/art/case-analysis-farc-colombia.

Tiempo, Redacción El. "La Ideología de Las Guerrillas." El Tiempo, October 14, 1995. https://www.eltiempo.com/archivo/documento/MAM-424318.

Tilly, Charles. *The Politics of Collective Violence*. Cambridge University Press, 2003.

"Timeline: America's War on Drugs." NPR, April 2, 2007. https://www.npr.org/templates/story/story.php?storyId=9252490.

"Timeline: U.s.-Colombia Relations." Council on Foreign Relations. Accessed September 24, 2023. https://www.cfr.org/timeline/us-colombia-relations.

"Transnational Organized Crime: A Growing Threat to National and International Security." National Archives and Records Administration. Accessed April 29, 2024. https://obamawhitehouse.archives.gov/administration/eop/nsc/transnational-crime/threat.

Tras Los Pasos Perdidos de la Guerra Sucia: Paramilitarismo y operaciones encubiertas en Colombia. Belgica: Ncos, 1995.

U.S. Army Special Operations Command (USASOC) History Department. “The 1960s.” The 1960s: A Decade of Revolution. Accessed September 12, 2023. https://arsof-history.org/articles/v4n4_1960s_page_1.html.

“Un Introduction to Organized Crime.” Essay. In *Organized Crime as Defined by Laws, Agencies, and Governments*. John Bartlett LLC, n.d.

“US Covert Program Exposed In Colombia.” The Prism, August 23, 1998. https://www.ibiblio.org/prism/sep98/uscovert.html.

Vidal Ortega, Antonino, and Jorge Enrique Elias Caro. The worker’s massacre of 1928 in the Magdalena zona banana - Colombia. An unfinished story, 2012. http://www.scielo.org.co/scielo.php?script=sci_abstract&pid=S1794-88862012000300003.

“Violencia Paramilitar En La Altillanura: Autodefensas ...” Centro Nacional de Memoria Historica, 2018. http://centrodememoriahistorica.gov.co/wp-content/uploads/2020/01/violencia-paramilitar-en-la-altillanura-1.pdf.

Walker, Jack L. A critique of the elitist theory of democracy, June 1966. https://fbaum.unc.edu/teaching/articles/Walker-APSR_1966.pdf.

“World Report 2023: Rights Trends in Colombia.” Human Rights Watch, January 20, 2023. https://www.hrw.org/world-report/2023/country-chapters/colombia.

“The World□: Colombian Judge Slain.” Los Angeles Times, October 31, 1986. https://www.latimes.com/archives/la-xpm-1986-10-31-mn-8353-story.html.

Wright, Eric Olin. "Analytical Marxism." *Encyclopedia of Social Theory*, 2005. https://doi.org/10.4135/9781412952552.n332.

Youngman, Mark. Ubira etheses - University of Birmingham e Data Repository. Accessed September 13, 2023. https://etheses.bham.ac.uk/id/eprint/9372/7/Youngman2019PhD.pdf.

Yumpu.com. "Fabio Ochoa-Vasquez Juan." yumpu.com. Accessed April 28, 2024. https://www.yumpu.com/en/document/read/20941454/download-this-file-the-free-information-society/78.

Ávila, Ariel. "Las Antiguas Zonas de Las FARC y La Disputa Por La Economía de Guerra." El País América Colombia, April 17, 2018. https://elpais.com/internacional/2018/04/17/colombia/1523997042_036762.html.

---

[1] Angel Rabasa and Peter Chalk. *Colombian Labyrinth the Synergy of Drugs and Insurgency and Its Implications for Regional Stability*. Santa Monica, CA: RAND, 2001.

[2] *Tras Los Pasos Perdidos De La Guerra Sucia: Paramilitarismo y Operaciones Encubiertas En Colombia*. Bruselas: NCOS, 1995

[3] On the agrarian origins of civil conflict in Colombia María del Pilar López-Uribe - London School of Economics and Political Science Fabio Sanchez Torres - Universidad de los Andes

[4] "Covert Action in Colombia." The Washington Post. WP Company. Accessed February 20, 2022.

[5] Noam Chomsky. Rogue states: The rule of force in World Affairs. London: Pluto Press, 2016.

[6] Alonso Salazar. *La Parabola de Pablo*. DEBOLSILLO, 2019.

[7] Alan Seagrave. Conflict in Colombia: How Can Rebel Forces, Paramilitary Groups, Drug Traffickers, and Government Forces Be Held Liable for Human Rights Violations in a Country Where Impunity Reigns Supreme?

[8] Ibid

[9] Colombia All-Groups, Ospina Exports to World in US$ Thousand 1988-1994 p.129

[10] Administrative Department of Security.
[11] Colombia's illegal armed groups (maps). Colombia reports
[12] Paramilitary violations of International Humanitarian Law.
[13] Charles Bergquist. *Violence in Colombia 1990-2000: Waging War and Negotiating Peace.* Wilmington, DE: SR Books, 2001. Print.
[14] Ana Arjona. *Rebelocracy: social order in the Colombian civil war.* Cambridge, U.K: Cambridge University Press, 2017.
[15] Associated Press. "Judge Investigating Escobar Is Shot to Death in Colombia (Published 1992)." The New York Times, September 19, 1992.
[16] "The World: Colombian Judge Slain." Los Angeles Times, October 31, 1986. https://www.latimes.com/archives/la-xpm-1986-10-31-mn-8353-story.html.
[17]Ana Arjona. *Rebelocracy: social order in the Colombian civil war.* Cambridge, U.K: Cambridge University Press, 2017.
[18] ibid
[19] Colombia Reports. "Cali Cartel Gave $10m to Samper's Presidential Campaign: Cartel Executive." Colombia News | Colombia Reports, February 26, 2015.
[20] Ibid Samper
[21] Ana Arjona. *Rebelocracy: social order in the Colombian civil war.* Cambridge, U.K: Cambridge University Press, 2017.
[22]Angel Rabasa and Peter Chalk. *Colombian labyrinth the Synergy of drugs and insurgency and its implications for regional stability.* Santa Monica, CA: RAND, 2001.
"*The implications of the U.S. assistance program for Colombia are that the United States, by taking the lead in mustering international support for Plan Colombia, has increased its political stake in Colombia. If the Pastrana administration falters, either in its counter-narcotics or counter-insurgency approach, the United States would be confronted with an unpalatable choice. It could escalate its commitment, to include perhaps an operational role for U.S. forces in Colombia, or scale it down, which could involve some significant costs, including a serious loss of credibility and degradation of the U.S. ability to muster regional support for its counter-narcotics and political objectives. At the same time, Bogotá has essentially adopted a U.S. vision of the problem—which focuses on narcotics eradication—as the centerpiece of its strategy. As a result, it has gained access to substantial U.S. resources, but at the cost of a loss of flexibility to design and implement its own solutions. P 53-70*
[23] Mark Moyar, Hector Pagan, and Wil R Griego. Persistent Engagement in Colom BIA - Washington Office on ..., July 2014. https://www.wola.org/files/141113_jsou_notes.pdf.
Although the Colombian military provided support to the paramilitaries, the extent and nature of the support were often complicated to prove or disprove. According to diligent U.S. embassy investigations, some of the accusations of military complicity in human rights violations proved to be unfounded, while others were found to contain truth.

[24] Covert action in Colombia | The Washington Post. Accessed April 16, 2024. https://www.washingtonpost.com/sf/investigative/2013/12/21/covert-action-in-colombia/.
[25] ibid
[26] ibid
[27] ibid
[28] ibid
[29] ibid
[30] ibid
[31] ibid
[32] ibid
[33] ibid
[34] "World Report 2023: Rights Trends in Colombia." Human Rights Watch, January 20, 2023. https://www.hrw.org/world-report/2023/country-chapters/colombia.
[35] Patric Dunleavyk, and Brendan O'Leary. *Theories of the state: The Politics of Liberal Democracy*. Basingstoke: Macmillan Education, 1987.
[36] Claudia Mariotti. "Elite Theory." The Palgrave Encyclopedia of Interest Groups, Lobbying and Public Affairs, December 9, 2020.
[37] ibid
[38] ibid
[39] ibid
[40] ibid
[41] ibid
[42] ibid
[43] ibid
[44] Marco Palacios. *Between legitimacy and violence: A history of colombia, 1875-2002*. Durham: Duke University Press, 2007.
[45] *"La historia política colombiana se caracteriza por la persistencia de un arraigado particularismo localista que se originó en la sociedad colonial y que, desde los albores del periodo "nacional" se ha considerado como uno de los obstáculos más formidables que se interponen en el proceso de la centralización política y la integración nacional."*
La fragmentación regional de las clases dominantes en Colombia: una perspectiva histórica MARCO PALACIOS
[46] "Department of Justice: Homepage: United States Department of Justice." Department of Justice | Homepage | United States Department of Justice, September 9, 2016. https://www.justice.gov/.
[47] The Perversion of Land Reform by Landed Elites: Power, Inequality and Development in Colombia
Paul Faguet et alt "The Perversion of Public Land Distribution by Landed Elites: Power, Inequality and Development in Colombia." World Development, August 4, 2020.
[48] Juan F Roman. "Guerrilla Violence in Colombia: Examining Causes and Consequences." Calhoun Home, June 1, 1994.
[49] Juan F Roman
(Direct Quote) The last civil war, the War of the Thousand Days (1899-1902), took more than 100,000 lives, ruined the capacity for production,

and allowed the separation of Panama by focusing the government on internal, rather than external considerations. At the end of this conflict, the elite inaugurated a Period of Reconciliation (1903-30), catalyzed by Rafael Reyes' presidency. The replacement of discredited extremist factions with a more conciliatory moderate faction made it possible for the two parties to share power and achieve a consensus on what policies were appropriate for Colombian society

[50]Hoyos - Zuñiga, Edgar Jesus. "Lucha Por La Tierra En El Resguardo Indigena Nasa de Las Delicias 1977-1986."

(Direct quote in Spanish) *Dada esta situación de abundancia de tierras y de escasez de trabajo, los empresarios agrícolas naturalmente buscaban ejercer control sobre territorios ya ocupados por colonos para generar una fuerza de trabajo dependiente. ¡Sólo restringiendo el libre acceso de los campesinos a las tierras públicas mejor localizada y, por tanto, privándolos de una alternativa económica, esperaban los empresarios poder sujetar a los campesinos a sus propiedades*! (p 23)

[51] Marx, Karl. "Chapter 26 The Secret of Primitive Accumulation." *Capital*, April 17, 2008.

[52] Ospina, Hernando Calvo. *Colombia, Laboratorio de embrujos democracia Y terrorismo de estado*. Tres Cantos, Madrid: Foca, 2008.

[53] Antonino Vidal Ortega , and Jorge Enrique Elias Caro. The worker's massacre of 1928 in the magdalena zona bananera - colombia. an unfinished story, 2012.

[54] Colombian President in 1934, Member of Colombian Liberal Passed Law 200

[55]"Las Farc-EP En El Caribe Colombiano." *El otro derecho*, November 10, 2020, 27–33.

[56] Marx, Karl. "Chapter 26 The Secret of Primitive Accumulation." *Capital*, April 17, 2008.

[57] Juan F Roman. "Guerrilla Violence in Colombia: Examining Causes and Consequences." Calhoun Home, June 1, 1994. https://calhoun.nps.edu/handle/10945/42913.

[58] Term used by Marx in his Capital. Ch 36. Primitive accumulation.

[59] Volatile Social Movement and the Origins of Terrorism. The radicalization of change. Christine Sixta Rinehart. P 78

[60] Ibid

[61] Colombian President

[62] Daniel Bonilla. "Liberalism and Property in Colombia: Property as a Right and Property as a Social Function." *Fordham Law Review* 80, no. 3 (December 2011).

[63]Ted Piccone. The Colombian experience with Transitional Justice, July 2019.

[64] Maria del Pilar Lopez Uribe, and Fabio Sanchez-Torres . On the agrarian origins of civil conflict in Colombia. Accessed April 12, 2024.

[65] Ibid

[66] Ted Piccone. The Colombian experience with Transitional Justice, July 2019.

[67] Francisco Gutierrez Sanín . "Tough Tradeoffs: Coca Crops and Agrarian Alternatives in Colombia." International Journal of Drug Policy, February 18, 2021.
[68] Christopher Cramer and Elisabeth Jean Wood. "Introduction: Land Rights, Restitution, Politics, and War in Colombia." *Journal of Agrarian Change* 17, no. 4 (September 20, 2017): 733–38.
[69] Francisco Gutierrez Sanín, and Mauricio Baron . Re-stating the state in Colombia - LSE, 2005. https://www.lse.ac.uk/international-development/Assets/Documents/PDFs/csrc-working-papers-phase-one/wp66-restating-the-state-in-colombia.pdf.
[70] Valbona karanxha saber and scroll
[71] Eric Olin Wright. "Analytical Marxism." *Encyclopedia of Social Theory*, 2005.
[72] Ibid
[73] ibid
[74] Ibid
[75] Brian James DeMar. *Land wars: The story of China's Agrarian Revolution*. Stanford, CA: Stanford University Press, 2019.
[76] ibid
[77] "PENSAMIENTO MARXISTA COLOMBIANO Y ORIGEN Y PERMANENCIA DE LAS FARC-EP." usergioarboleda.edu, January 2016.
[78] Charles Tilly. *The Politics of Collective Violence*. Cambridge University Press, 2003.
[79] Guide to the analysis of insurgency, January 5, 2009. https://irp.fas.org/cia/product/insurgency.pdf.
[80] Ibid
[81] Ibid
[82] Ibid
[83]Charles Tilly "From mobilization to revolution." University of Michigan. 1977. P.7-4 Chapter Seven Revolution and rebellion.
[84] El Tiempo, Redacción . "La Ideología de Las Guerrillas." El Tiempo, October 14, 1995.
[85] "Banditry and Insurgency in Colombia." CIA.GOV, June 2001.
[86] Ariel Ávila, "Las Antiguas Zonas de Las FARC y La Disputa Por La Economía de Guerra." El País América Colombia, April 17, 2018. https://elpais.com/internacional/2018/04/17/colombia/1523997042_036762.html.
(Direct Quote*) Las FARC operaban en 242 municipios del país, lo que corresponde a un 22% del total de municipios de Colombia. Estas zonas, en su mayoría, se caracterizaban por tres cosas. Por un lado, eran zonas de frontera y municipios categoría seis, es decir, con poca población y muy pobres. En segundo lugar, son zonas con un gran déficit en materia de construcción de Estado. Es decir, territorios con grados bajos de* integración al mercado nacional, ausencia de instituciones básicas del Estado y un sistema de regulación de la vida social que ejercieron las FARC, como la administración de justicia, principalmente en las zonas rurales de estos municipios.
[87] Charles Tilly 7-12

[88] Ibid.
[89] BANDITRY AND INSURGENCY IN COLOMBIA - The World Factbook. Accessed September 20, 2023.
[90] ibid
[91] Ibid
[92] Ejercito Popular de Liberacion , and Ejercito de Liberacion Nacional (Popular Liberation Army, National Liberation Army)
[93] Garry Leech. *The FARC: The longest insurgency*. London: Zed Books Ltd, 2011.
[94] *Partido Communista Colombiano (Colombian Communist Party). Pedro Antonio Marin was one of the movement autodefensa leaders. Later he changed his name to Manuel Marulanda Velez and will become the supreme commander op FARC. Manuel Marulanda Velez was born in 1930 and died in 2008.*
[95] Garry Leech. *The FARC: The longest insurgency*. London: Zed Books Ltd, 2011. p 13
[96] Paul Collier, and Anke Hoeffler. Greed and Grievance in Civil War.
[97] Angel Rabasa, and Peter Chalk. *Colombian labyrinth the Synergy of drugs and insurgency and its implications for regional stability*. Santa Monica, CA: RAND, 2001.
[98] Ibid
[99] Kevin A Self. COUNTERTERRORISM POLICY IN COLOMBIA. Accessed September 23, 2023. https://apps.dtic.mil/sti/pdfs/ADA389095.pdf.
[100] Nash Jenkins. "Colombia: More than 100 Butchered Bodies Found beneath Jail." Time, February 19, 2016
[101] Claudia Hofmann and Urlich Schneckener . "Engaging Non-State Armed Actors in Stateand Peace-Building: Options and Strategies." *International Review of the Red Cross* 93, no. 883 (September 2011).
[102] Ibid
[103] ibid
[104] Ibid
[105] *Tras Los Pasos Perdidos de la Guerra Sucia: Paramilitarismo y operaciones encubiertas en Colombia*. Belgica: Ncos, 1995.
[106]ibid
[107] ibid
[108] "Violencia Paramilitar En La Altillanura: Autodefensas ..." Centro Nacional de Memoria Historica, 2018.
Bloque Centauros, las Autodefensas Campesinas del Casanare, el Bloque Vencedores de Arauca, el Frente Vichada del Bloque Central Bolívar y las Autodefensas Campesinas de Meta y Vichada.
[109] *Tras Los Pasos Perdidos de la Guerra Sucia: Paramilitarismo y operaciones encubiertas en Colombia*. Belgica: Ncos, 1995.
[110] ibid
[111] Ibid. En 1987,el general Alvaro ValenciaTovar concluía que "el Ejército demanda de la- sociedad actitudes solidarias cuando trata de defenderla de sus enemigos... (que) interesa lograr una-plenasimbiosis entreíos instrumentos armados de la nación y el pueblo que los nutre

con sus hombres, los apoya con su solidariday debe integrarse con ellos para formar un frente sólido, identificado ante la amenaza... (y que) la solidaridad civil-militar debe conseguirse con la amplitud requerida para rechazar la agresión donde se haga evidente y preparar de- . tensas adecuadas en los sectores más vulnerables. El Espectador, 2 de dic~mbre de 1991:pág. 11 A. Cited in tras los pasos perdidos de la guerra sucia.

[112] And particularly General Alvaro Valencia Tovar and Colonel l Orlando Zafra Galvis (The elite grappling to hold on to power)

[113] *Los agentes clandestinos y los. agitadores se infiltran en las organizaciones sociales y estatales, se apoderan de los cargos directivos y... generan planes de debilitamiento de las estructuras con miras a crear el caos y el derrumbamiento final del Estado. No existe instituto, organización o grupo social, político o religioso, sobre el cual no tengan interés de penetrarlo y dominarlo. Todas estas actividades... constituyen lo que se denomina la guerra política que., es la parte más peligrosa para la vida de las democracias' (pg 11-14)*

[114] Revista de las fuerzas armadas. *Tras Los Pasos Perdidos de la Guerra Sucia: Paramilitarismo y operaciones encubiertas en Colombia*. Belgica: Ncos, 1995.

[115] Brett, Roddy. "Elite Bargains and Political Deals Project: Colombia Case Stud." Stabilization Unit, February 2018.

[116] ibid

[117] Roddy Brett. "Elite Bargains and Political Deals Project: Colombia Case Stud." Stabilization Unit, February 2018.

By the end of the 1980s, the FARC-EP's strategy began to represent a threat to the economic status quo, as the insurgency began increasingly to assert control in rural areas. Consequently, an intense wave of paramilitary violence, sponsored by the state and the economic elite, began in the 1990s in response to the country's insurgencies. This paramilitary violence drew upon the presence of private militia groups that had evolved over a number of years, culminating in 1997 in the establishment of the United Self-Defence Forces of Colombia (AUC), which were closely linked to the formal political elite and economic actors.8 By the mid-2000s the AUC's violent response weakened the insurgency in key conflict areas, pushing the guerrillas back to peripheral zones

[118] Paramilitarismo Balance de la contribución del CNMH al esclarecimiento histórico, 2018.

[119]Archila, Mauricio. "Violencia y Política En Colombia. De La Nación Fragmentada a La Contrucción Del Estado | Anuario Colombiano de Historia Social y de La Cultura." repositorio.unal.edu, 2003.
[120] Ibid 53-55
[121] ibid
[122] ibid
[123] Paramilitarismo Balance de la contribución del CNMH al esclarecimiento histórico, 2018. (pg 19)
[124] ibid
[125] Paramilitarismo Balance de la contribución del CNMH al esclarecimiento histórico, 2018.
El principal cambio del replanteamiento estratégico del paramilitarismo consistió en priorizar la lucha por el territorio y la proyección política. Del paramilitarismo expedicionario, punitivo y fragmentado de los años ochenta se pasa a un paramilitarismo coordinado, territorializado y politizado (CNMH, 2016, página 139)
[126] Restrepo, José Alvear. "The Contribution of Chiquita Corporate Officials to Crimes against Humanity in Colombia." International Human Right Clinic, 2017.
[127] ibid
[128] ibid
[129] *Tras Los Pasos Perdidos de la Guerra Sucia: Paramilitarismo y operaciones encubiertas en Colombia*. Belgica: Ncos, 1995. Pg 36
[130] *Tras Los Pasos Perdidos de la Guerra Sucia: Paramilitarismo y operaciones encubiertas en Colombia*. Belgica: Ncos, 1995.
CODINAL Curso integral de defensa nacional- C was run by the Escuela Superior de La Guerra (Superior War/Combat School) one a year where it recruited people from the public and private sector.
Also, the military offered course for a reserve force where professionals from the Higher Education can receive training without quiting their jobs (Decree 613 de 1977 y 89 de 1984) pg 36-67
[131] *On 16 May, the paramilitaries staged a daring attack on the river port of Barrancabermeja, killing 11 people and kidnapping more than 20, others; the kidnap victims were subsequently executed. The audacity of the attack, which occurred in an area with a heavy guerrilla presence and was forewarned a year earlier, left many wondering whether Colombia is on the verge of another period of intense paramilitary violence against guerrillas and their alleged sympathizers*
[132] "The Kingpin Strategy: Did It Work and Is It Still Relevant?" The Kingpin Strategy: Did It Work and Is It Still Relevant? | DEA Museum.
[133] "Transnational Organized Crime: A Growing Threat to National and International Security." National Archives and Records Administration.
[134] Hillary Higgins. "Counternarcotics to Counterinsurgency: Assessing the Transformation of US Economic Assistance to Colombia, 1998–2002.
[135] Ibid

[136] Tom Harvey. "Colombia Close to Capturing 'public Enemy No. 1' - UPI Archives." UPI, May 20, 1990.
[137] ibid
[138]*HEARINGS BEFORE T.rIE PERMANENT SUBCOMMITTEE ON INVESTIGATIONS OF THE COMMrrTEE ON GOVERNMENTAL AFFAIRS UNITED STATES SENATE ONE HUNDRED FIRST CONGRESS, 1989.*
[139] Tom Harvey. "Colombia Close to Capturing 'public Enemy No. 1' - UPI Archives." UPI, May 20, 1990.
[140] Bloque Busqueda. Joint police force created to combat the drug cartel
[141] Kingpin strategy.
[142] Tom Harvey. "Colombia Close to Capturing 'public Enemy No. 1' - UPI Archives." UPI, May 20, 1990.
[143] ibid
[144] It had an arsenal of weapons:
[145] "Timeline: America's War on Drugs." NPR, April 2, 2007.
[146] El patreon del mal SALAZAR, ALONSO. *La Parabola de Pablo*. DEBOLSILLO, 2019.
[147]Ibid (Salazar)
[148] "Un Introduction to Organized Crime." Essay. In Organized Crime as Defined by Laws, Agencies, and Governments. John Bartlett LLC, n.d.
[149] "The Drug Enforcement Administration (DEA) 1980-1985." The Drug Enforcement Administration. Accessed April 26, 2024.
[150] ibid
[151] ibid
[152] ibid
By 1985, Colombia had the highest murder rate in the world.
[153] https://colombiareports.com/pablo-escobars-involvement-in-1985-palace-of-justice-attack-another-urban-myth/
[154] Castaño, Carlos, and Mauricio Aranguren Molina. *Mi Confesión: Carlos Castaño Revela Sus Secretos*. Bogotá: Oveja Negra, 2002.
[155] SALAZAR, ALONSO. *La Parabola de Pablo*. DEBOLSILLO, 2019.
[156] Carlos Enrique Lehder Rivas was a German-Colombian former drug lord who was co-founder of the Medellín Cartel. He was released from prison in the United States after 33 years in 2020.
[157]Defense Intelligence Agency > FOIA > Foia Electronic Reading Room > Foia Reading Room: Colombia. Accessed April 27, 2024.
[158] The Drug Enforcement Administration (DEA) 1990-1994. Accessed April 24, 2024.
[159] Ibid
[160] The Drug Enforcement Administration (DEA) 1990-1994. Accessed April 24, 2024.
Fabio Ochoa surrendered to authorities near Medellin. Shortly, after, in January 1991, Fabio Ochoa's brother, Jorge Luis, also turned himself into the CNP. The brothers, along with Pablo Escobar, had been the top leaders of the cartel. In addition, in January 1991, the CNP killed David Ricardo Prisco Lopera, Pablo Escobar's top assassin, along with his younger brother, Armando Alberto Prisco.

[161] Los Priscos was a group of criminals at the service of the Medellín cartel that participated in the armed conflict in Colombia in the 1980s and the beginning of the following decade. He has been accused of participating in several of the assassinations perpetrated in Colombia. At its peak, the organization recruited more than 300 men to carry out kidnappings, terrorist attacks and murders. (Wiki Spanish)
[162] The Drug Enforcement Administration (DEA) 1990-1994. Accessed April 24, 2024.
[163] ibid
[164]. "Fabio Ochoa-Vasquez Juan." yumpu.com. Accessed April 28, 2024.
[165] The Drug Enforcement Administration (DEA) 1990-1994. Accessed April 24, 2024. https://www.dea.gov/sites/default/files/2021-04/1990-1994_p_67-76.pdf.
[166] Fulo Regilme Jr, Salvador Santino. "American Foreign Aid and Colombia's Human Rights Tragedy." E International Relations, July 7, 2018.
[167] Briscoe, Charles H. "Across the Pacific to War." Across the Pacific to War: The Colombian Navy in Korea, 1951–1955, 2006.
[168] Ibid
[169] Richburgh , Keith B. "Reagan Order Defines Drug Trade as Security Threat, Widens Military Role - The Washington Post." Washington Post , 1986.
[170] Joe Biden. " AID TO ``PLAN COLOMBIA'': THE TIME FOR U.S. ASSISTANCE IS NOW."
[171] DeYoung, Karen. "Politics - The Washington Post." Washington Post , August 31, 2000.
[172] Miroff, Nick. 'Plan Colombia': How washington learned to love Latin American intervention again - The Washington Post, September 6, 2006.
[173] ibid
[174] Covert action in Colombia | The Washington Post. Accessed April 16, 2024.
[175] ibid
[176] US Covert Program in Colombia
"Some of the personnel involved in US operations in Colombia are working under a State Department contract with Dyncorp and East Inc., private firms based in suburban Washington.
[177] The neoliberal party was a proponent supporter of the extradition law and it promised a cooperation between the US and Colombian judicial system.
[178] Hartlyn, Jonathan. "Drug Trafficking and Democracy in Colombia in the 1980s ..."
The extradition treaty fell in a complex and judicially dubious process and attempts to revive it as of mid-1989 were unsuccessful. In December 1986, the Supreme Court declared the extradition treaty with the United States null and void because it had not been signed by the President (who had been out of the country) but by the cabinet minister who had been acting as president. In June 1987, the Supreme Court declared unconstitutional the effort of President Barco to sanction the treaty by a new law following the first

finding. Efforts to employ an earlier treaty as a basis for extradition were rejected in April and May 1988 by the Council of State.
[179]DRUG ENFORCEMENT ADMINISTRATION 1990-1994, 1994.
[180] ibid"Colombia CJA Litigation Cabrera v. Jiménez Naranjo Urán, et al. v. Plazas Vega." CJA. Accessed April 24, 2024. https://cja.org/where-we-work/colombia/.
[181] ibid
[182] ibid
[183] ibid
[184] M-19 guerrilla attack on Colombian Palace of ... Accessed April 24, 2024. https://www.cia.gov/readingroom/docs/CIA-RDP91B00874R000100040006-2.pdf.
[185] "The Palace of Justice Attack – Losses and Gains."
[186] Ibid.
[187] *El Palacio de Justicia fue una historia al desnudo, una representación teatral en espacio público en la que los actores se sobreactuaron natural y espontáneamente como si nadie los estuviese mirando (...) Que todo se hacía para "defender la democracia, maestro", dijo uno; que "todavía hay hombres dispuestos a morir por una causa" coreó el otro. Y el país, atónito, los escuchó (Vélez y Atehortúa, 1993: 291).*
[188] M 19 is known as the April 19 Movement; a guerrilla group who operated in Colombia during the 1980s and 1990s. They were a Marxist ideological political organization with mostly urban intellectuals who pressured the government by kidnapping officials, diplomats. The most notorious remains the Siege of Palacio de Justicia to trial former Colombian President Betancourt
[189]Francisco, Alvaro, Amaya Villarreal, and Gustavo Emilio Cote-Barco. Vista de la Toma del Palacio de Justicia: La reparación del daño en eventos de violación de derechos humanos, Winter 2006
[190] *DAS*
[191] *Carlos Pizarro was a demobilized member of M19 who had signed an agreement with the state of Colombia to surrender the arms and stop all subversive activities was assassinated 47 days after the peace agreement while he had declared his candidacy for president of Colombia. Carlos Pizarro Leongómez (6 June 1951 – 26 April 1990) was the fourth commander of the Colombian guerrilla group 19 of the April Movement.. Pizarro later ran for president of Colombia after the demobilization of M-19* that transformed the group into the political party. M-19 Democratic Alliance (*Alianza Democrática M-19*) (AD/M-19). Pizarro was assassinated on 26 April 1990 by the paramilitary.
[192] "An Overview of Conflict in Colombia." International Center for Transitional Justice, 2009.
[193] Piccone, Ted. The Colombian experience with Transitional Justice, July 2019.
[194] ibid
[195]Juan Papier. "Left Undefended." Human Rights Watch, August 17, 2023
According to a February 26 report from the Office of the United Nations High Commissioner for Human Rights (OHCHR), there were 108 verified

killings of social leaders and human rights defenders in 2019. According to the Attorney General's Office, in the cases of more than 400 killings of human rights defenders from January 2016 to August 2020, the government had obtained 60 convictions. According to the OHCHR, 75 percent of the 2019 social leader killings occurred in rural areas, and 98 percent occurred in areas where the ELN and other criminal groups were present. The motives for the killings varied, and it was often difficult to determine the primary or precise motive in individual cases. For example, on March 19, armed men reportedly kidnapped and killed crop substitution activist Marco Rivadeneira in Puerto Asis, Putumayo. On April 10, authorities arrested Abel Antonio Loaiza Quinonez, alias "Azul," in Puerto Asis. According to officials in the Attorney General's Office, Azul was a senior member of an illegal armed group linked to several killings in the region, possibly including the killing of Rivadeneira.

[196] HUMAN-RIGHTS-REPORT.pdf

[197] Hannah Stone. "Colombia Elites and Organized Crime: Introduction." InSight Crime, October 22, 2023.

[198] Jack L. Walker. A critique of the elitist theory of democracy, June 1966.

[199] ibid

Made in the USA
Monee, IL
04 November 2026

67377456R00105